AF736866

AFRICAN AMERICANS AND THE MISSISSIPPI RIVER

This book follows the historical trajectory of African Americans and their relationship with the Mississippi River dating back to the 1700s and ending with Hurricane Katrina and the still-contested Delta landscape.

Long touted in literary and historical works, the Mississippi River remains an iconic presence in the American landscape. Whether referred to as "Old Man River" or the "Big Muddy," the Mississippi River represents imageries ranging from the pastoral and Acadian to turbulent and unpredictable. However, these imageries—revealed through the cultural production of artists, writers, poets, musicians, and even filmmakers—did not reflect the experiences of everyone living and working along the river. Missing is a broader discourse of the African American community and the Mississippi River. Through the experiences of African Americans with the Mississippi River, which included narratives of labor (free and enslaved), refuge, floods, and migration, a different history of the river and its environs emerges.

The book brings multiple perspectives together to explore this rich history of the Mississippi River through the intersection of race and class with the environment. The text will be of great interest to students and researchers in environmental humanities, including environmental justice studies, ethnic studies, and US and African American history.

Dorothy Zeisler-Vralsted is Professor Emeritus at Eastern Washington University, USA. Her publications include *Rivers, Memory and Nation-Building: A History of the Volga and Mississippi Rivers*, and *Water and Human Societies: Historical and Contemporary Perspectives*, coedited with David A. Pietz.

Routledge Environmental Humanities

Series editors: **Scott Slovic** (University of Idaho, USA),
Joni Adamson (Arizona State University, USA) and
Yuki Masami (Aoyama Gakuin University, Japan)

Editorial Board

Christina Alt, St Andrews University, UK
Alison Bashford, University of New South Wales, Australia
Peter Coates, University of Bristol, UK
Thom van Dooren, University of Sydney, Australia
Georgina Endfield, Liverpool, UK
Jodi Frawley, University of Western Australia, Australia
Andrea Gaynor, The University of Western Australia, Australia
Christina Gerhardt, University of Hawai'i at Mānoa, USA
Tom Lynch, University of Nebraska, Lincoln, USA
Iain McCalman, Australian Catholic University, Australia
Jennifer Newell, Australian Museum, Sydney, Australia
Simon Pooley, Imperial College London, UK
Sandra Swart, Stellenbosch University, South Africa
Ann Waltner, University of Minnesota, US
Jessica Weir, University of Western Sydney, Australia

International Advisory Board

William Beinart, University of Oxford, UK
Jane Carruthers, University of South Africa, Pretoria, South Africa
Dipesh Chakrabarty, University of Chicago, USA
Paul Holm, Trinity College, Dublin, Republic of Ireland
Shen Hou, Renmin University of China, Beijing, China
Rob Nixon, Princeton University, Princeton NJ, USA
Pauline Phemister, Institute of Advanced Studies in the Humanities, University of Edinburgh, UK

Sverker Sorlin, KTH Environmental Humanities Laboratory, Royal Institute of Technology, Stockholm, Sweden
Helmuth Trischler, Deutsches Museum, Munich and Co-Director, Rachel Carson Centre, Ludwig-Maximilians-Universität, Germany
Mary Evelyn Tucker, Yale University, USA
Kirsten Wehner, University of London, UK

The *Routledge Environmental Humanities* series is an original and inspiring venture recognising that today's world agricultural and water crises, ocean pollution and resource depletion, global warming from greenhouse gases, urban sprawl, over-population, food insecurity and environmental justice are all *crises of culture.*

The reality of understanding and finding adaptive solutions to our present and future environmental challenges has shifted the epicenter of environmental studies away from an exclusively scientific and technological framework to one that depends on the human-focused disciplines and ideas of the humanities and allied social sciences.

We thus welcome book proposals from all humanities and social sciences disciplines for an inclusive and interdisciplinary series. We favour manuscripts aimed at an international readership and written in a lively and accessible style. The readership comprises scholars and students from the humanities and social sciences and thoughtful readers concerned about the human dimensions of environmental change.

For more information about this series, please visit: www.routledge.com/Routledge-Environmental-Humanities/book-series/REH

AFRICAN AMERICANS AND THE MISSISSIPPI RIVER

Race, History, and the Environment

Dorothy Zeisler-Vralsted

Routledge
Taylor & Francis Group
LONDON AND NEW YORK

earthscan
from Routledge

Cover image: © "Watching the sternwheel steamboat pass"; Earl S. Miers River Photo Collection, Courtesy of the Tennessee State Library and Archives

First published 2023
by Routledge
4 Park Square, Milton Park, Abingdon, Oxon OX14 4RN

and by Routledge
605 Third Avenue, New York, NY 10158

Routledge is an imprint of the Taylor & Francis Group, an informa business

© 2023 Dorothy Zeisler-Vralsted

The right of Dorothy Zeisler-Vralsted to be identified as author of this work has been asserted in accordance with sections 77 and 78 of the Copyright, Designs and Patents Act 1988.

All rights reserved. No part of this book may be reprinted or reproduced or utilised in any form or by any electronic, mechanical, or other means, now known or hereafter invented, including photocopying and recording, or in any information storage or retrieval system, without permission in writing from the publishers.

Trademark notice: Product or corporate names may be trademarks or registered trademarks, and are used only for identification and explanation without intent to infringe.

British Library Cataloguing-in-Publication Data
A catalogue record for this book is available from the British Library

Library of Congress Cataloging-in-Publication Data
Names: Zeisler-Vralsted, Dorothy, author.
Title: African Americans and the Mississippi River: race, history, and the environment / Dorothy Zeisler-Vralsted.
Description: London; New York, NY: Routledge, Taylor & Francis Group, 2023. | Series: Routledge Environmental Humantities | Includes bibliographical references and index.
Identifiers: LCCN 2022015251 (print) | LCCN 2022015252 (ebook) | ISBN 9781138671324 (hardback) | ISBN 9781138671331 (paperback) | ISBN 9781315617077 (ebook)
Subjects: LCSH: African Americans—Mississippi River Valley—History. | Mississippi River—History. | Mississippi River Valley—History. | Mississippi River—Environmental conditions. | Mississippi River Valley—Environmental conditions.
Classification: LCC F358.2.B53 Z45 2023 (print) | LCC F358.2.B53 (ebook) | DDC 977.004/96073—dc23/eng/20220414
LC record available at https://lccn.loc.gov/2022015251
LC ebook record available at https://lccn.loc.gov/2022015252

ISBN: 978-1-138-67132-4 (hbk)
ISBN: 978-1-138-67133-1 (pbk)
ISBN: 978-1-315-61707-7 (ebk)

DOI: 10.4324/9781315617077

Typeset in Bembo
by codeMantra

CONTENTS

FIGURES

PREFACE

Twenty years ago, I would not have written this book. For most of my academic and professional life, my interest in rivers focused upon humans' manipulations of rivers whether through dams, irrigation canals, or hydropower facilities. But when researching the Volga and Mississippi Rivers for a comparative study of major rivers, with their similar roles in fostering culture, nationalism, and each country's economic well-being, I recognized another commonality between the two. Exploring each river through the lens of labor, it soon became apparent how those who worked on the river, whether the Volga burlaki (barge haulers) or African American roustabouts, experienced the river in ways distinct from other groups. Their experiences, in turn, informed rich subcultures, as witnessed through song, folklore, and poetry.

Thus, for those laboring on the river, whether hauling a barge upstream or fueling the boilers of a Mississippi steamboat, each river provided livelihoods in an intimate exchange of labor and reward. The rivers did more than sustain, as linkages influenced cultural expressions through song or folklore that either celebrated or lamented the river, recounted exploits with the river, or identified the river as a means to freedom. For African Americans, enslaved for almost three centuries, the river as freedom included a possible escape to the surrounding cypress swamps of the Lower Mississippi or freedom upriver to the northern states. In contrast, the same river as highway could also mean a Second Middle Passage, as the trip downstream to the Deep South ensured enslavement in a brutal slave regime. These representations of the river differed significantly from the dominant narrative of the Mississippi.

In contrast, for the first Euro-Americans arriving in the Lower Mississippi Valley, the river represented an outlet to commerce—cementing the rise to empire—and eventually to a global economy where cotton reigned. Lured by handsome profits, colonists and later American planters viewed the river as a resource to be subdued, guaranteeing the safe passage of goods. But the Mississippi River also paralleled the history of the early United States, and an antebellum narrative that framed the river as integral to American exceptionalism. Poets, such as Henry Wadsworth Longfellow, and artists, particularly George Caleb Bingham, depicted a river that was part of an emergent American nationalism, embodying all the nuances of the nineteenth-century United States. Thus, the river served as the boundary between the frontier and civilization, as the river allowed for the unconventionality of the flatboatmen, part of an Americana captured by Bingham in "The Jolly Flatboatmen." By the twentieth century, as engineers embraced the promise of technology with a missionary zeal, the river became a resource to be tamed and harnessed. Flood control became the mantra in the dominant narrative for white Southern boosters, developers, and planters. In contrast, when examining the responses of African Americans to the river in the early twentieth century, a different story surfaced. For example, the blues music characterizing the 1927 Flood spoke of destruction, homelessness, and flight. Coinciding with reactions to the flood were the songs that portrayed life working in the levee camps—the federal response to flood control. Again, the narratives differed as for those laboring in the camps, conditions were often little better than those in the antebellum period when enslaved. The perspective of flood control to control an unruly river, protecting homes and fostering commerce, was a long way off from the everyday lives of the levee workers.

In other words, the more I researched the river from the perspective of African Americans experiencing the Mississippi River, the more a different river emerged. And while numerous texts include parts of this history, the story of the river from the early 1700s when the first Africans were forcibly brought to the site of New Orleans up to the present has not been recorded. My hope is that the following text—addressing the history of the river from this perspective—becomes part of the river's history, familiar to everyone. Ultimately, a more inclusive Mississippi River history will result with all the actors and their associations with the river. In the future, when we think of "Old Man River," not only is the imagery of steamboats dotting the landscape part of the popular imagination but also accompanying the imagery is the river as a site of violence and trauma for African Americans. Another narrative will challenge the role of swamps and wetlands in U.S. history; thus, when looking at the cypress swamps bordering the Lower Mississippi River, our thoughts will turn to what these environs held for enslaved populations in a history rife with examples of resilience to first enslavement and later, the Jim Crow South. But my research also revealed that this is not a binary history as the dominant narrative often converged with African American perspectives. In both instances, the river represented strength and a capricious nature. In both instances, the beauty of the river was acknowledged. In both instances, the cruelty of a flood was expressed by all. Hopefully, this is the history that will be internalized and when tourists view the legacy of steamboat captains, for example, they will also recall that the same steamboats participated in the enslavement of millions.

That said, the following text is not without omissions. Since the focus was African American experiences with the Mississippi, other groups, particularly Native Americans who lived along the river, have been overlooked. The relationship, however, between the two in the early years of colonial settlement was noted but not explored in close detail. Further, for each era of North American history, beginning with the colonial period, the literature review was limited. Scholarly works on slavery, for example, are extensive and although many of the major works were cited, many texts were omitted. As the focus was on the Mississippi River as seen through African Americans' experiences, my research centered on primary sources such as the narratives of formally enslaved people, whenever possible. Songs and folklore also illustrated their experiences. Thus the following history is not a definitive one but supplements our understanding of the Mississippi River and the role it played informing cultures.

In writing this book, I have many people and organizations to thank. My initial research began with the 1927 Flood and the blues music that it inspired. A conference at Renmin University, sponsored by the Rachel Carson Center, provided a venue for this research. Participation in the Environmental Humanities Seminar Series at the KTH Royal Institution of Technology in Stockholm offered another opportunity to discuss ongoing research. Presenting in the seminar series also introduced me to the well-known environmental historian, Libby Robin, one of the former editors of the Routledge Environmental Humanities Series. Further insights were gained when speaking at the Watersheds Symposium, hosted by the University of La Trobe's Centre for the Study of the Inlands. At the symposium, I received critical feedback from numerous water history scholars and had the good fortune to meet Trevor Hogan, one of the editors of *Thesis Eleven*. Under Trevor's guidance, a collection of symposium presentations was assembled in a special edition in which I published an introduction to the present text, titled "African Americans and the Mississippi River: Race, History and the Environment," *Thesis Eleven* (2019). Another publication titled "Working Lives on the Mississippi and Volga Rivers Nineteenth Century Perspectives" in *The Review of International American Studies* (2021) allowed for further reflection on the connection between labor and major rivers. Adding to these opportunities were the suggestions given at several junctures by colleagues well versed in water history.

However, none of this could have been accomplished without the assistance of the staff at numerous libraries, historical societies, archives, and museums. These institutions include the Percy Library in Greenville, Mississippi with its impressive oral history collection chronicling the 1927 Flood, which served as the springboard for further research. The library at Delta State University along with visits to area blues museums aided in understanding blues culture and twentieth-century Delta history. Other museums and archives that lent further understanding from the colonial era up to the present include the Herman T. Pott National Inland Waterways Library at the University of Missouri-St. Louis, State Historical Society of Missouri, Tennessee State Archives, New Orleans Public

Library, and Historic New Orleans Collection. In many instances, not only did the staff assist with gathering materials but they also reproduced the images found in the text. In addition to on-site visits to collections were the numerous digital collections, cited in the bibliography. Travel to the sites was often funded by Eastern Washington University, whose library staff was tireless in helping me locate sources and countless interlibrary loans. Finally, special thanks are due to Routledge editorial assistants, Matthew Shobbrook and Rosie Anderson. Their patience, encouragement, and assistance was always appreciated.

On a personal note, so many colleagues and friends have offered not only advice but also words of encouragement when needed the most. However, as always, my family have been my strongest supporters, sharing with me their time and resources, which enabled me to complete this work.

Every effort has been made to contact copyright holders for their permission to reprint material in this book. The publishers would be grateful to hear from any copyright holder who is not acknowledged here and would undertake to rectify any errors or omissions in future editions of this book.

INTRODUCTION

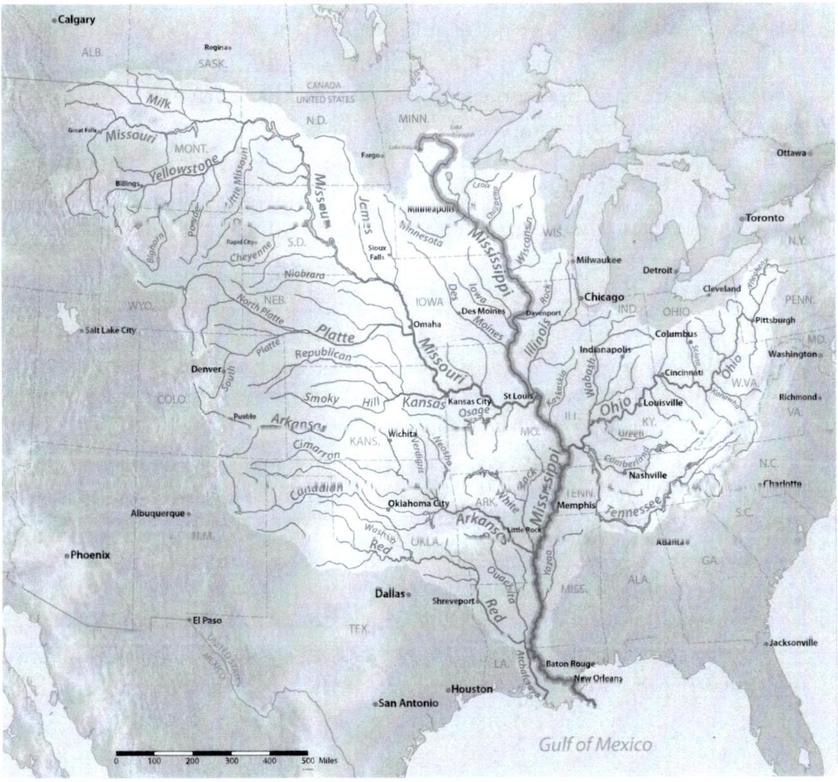

FIGURE 0.1 Mississippi River and Tributaries.
Credit: N/A.

DOI: 10.4324/9781315617077-1

> So many have written about the river . . . . I have read what most have written, and without them, I could not add another book to the list. Yet when you leave them you are still confused, for their eyes saw different things and at different times. The navigator tells his story and the explorer his, and the historian and the planter, the naturalist and the pilot and the soldier. The river is in all of them. But if you try to use them piecemeal, you find a patternless puzzle.
>
> *Lower Mississippi* (1942)
> Hodding Carter

Rivers have long been a subject in the popular imagination. For the Mississippi River (Figure 0.1), as the quote by Hodding Carter above illustrates, accounts vary from contemporary memoirs of journeys downriver to early nationalist accounts valorizing American exceptionalism to environmental histories, fusing the interplay of humans with their surroundings.[1] All these accounts draw from a rich repository of river histories chronicling the role that rivers have played in civilizations. Not one of these considers Egyptian civilization without the Nile, Mesopotamia without the Tigris and Euphrates, Russia without the Volga, England without the Thames, or China without the Yellow River, to name a few well-known examples. Many of the earlier histories are simple narratives, elaborating upon the river's place in a nation's cultural memories. In Russia, the Volga River was drafted into serving an emergent nineteenth-century nationalism, which was dependent upon distinguishing a riverine landscape that differed from the perceived more scenic European vistas. The Seine River in France has alternated from a utilitarian waterway, providing everything from laundry services to transportation, to its present renaissance as a scenic byway. In Roman civilization the Tiber River was not only commemorated in Roman mythology, but it also furnished water for drinking and agriculture as well as facilitated shipping. Finally, the Tiber became famous as the birthplace of Rome's founders: Romulus and Remus.[2]

The Mississippi River, given its length, strength, and sheer volume, was not exempt from the pageantry bestowed upon major rivers. Beginning with its name, which is derived from the Ojibwe "Misi-Ziibi" meaning "long river," North America's first inhabitants recognized the size of the river. Another Upper Midwestern tribe, the Winnebago, refers to the Mississippi as "Big Water" in their folklore. When Hernando De Soto, considered the first European to see the river, encountered the Mississippi River, he described it thus:

> as nearly a half league wide, and if a man stood still on the other side, one could not tell whether he were a man or something else. It [the river] was of great depth and of very strong current. Its water was always turgid and continually many trees and wood came down it borne along by the force of water and current.

As De Soto indicated, the river would also become known for its challenge to navigation, with early accounts lamenting a river with varying depths, the propensity to jump banks, and menacing snags. Still, by the time the French had reached the southern stretches of the Mississippi River and determined upon the site of New Orleans for a French outpost, the river was deemed indispensable for trade; a guarantee for a prosperous French colonial empire. The Mississippi remained a critical asset to the succeeding empires of Spain, Britain, and finally the United States.[3]

By the nineteenth century, the Mississippi River, as part of the Louisiana Purchase, was an integral part of popular imagery depicting the American landscape. For many, the river represented the beginning of the frontier, the demarcation between civilization and lawlessness. To others, the river was key to the nation's future in a physical and symbolic sense. But above all, the river would be one of the country's major arteries; the conduit for transporting goods from the interior to the Atlantic and a burgeoning global market. As the young nation's highway, the river was immortalized by those mesmerized by the idea of an American exceptionalism. Poets, artists, and writers celebrated the river and its role in shaping history. Henry Wadsworth Longfellow, the young nation's poet laureate, secured the river's place in a romantic American past with stanzas such as this found in the poem "Evangeline."

> It was the month of May. Far down the Beautiful River,
> Past the Ohio shore and past the mouth of the Wabash,
> Into the golden stream of the broad and swift Mississippi,
> Floated a cumbrous boat, that was rowed by Acadian boatmen.
> Day after day they glided adown the turbulent river;
> Night after night, by their blazing fires, encamped on its borders.
> Now through rushing chutes, among green islands, where plumelike
> Cotton-trees nodded their shadowy crests, they swept with the current,
> Then emerged into broad lagoons, where silvery sand-bars
> Lay in the stream, and along the wimpling waves of their margin,
> Shining with snow-white plumes, large flocks of pelicans waded.
> Level the landscape grew, and along the shores of the river,
> Shaded by china-trees, in the midst of luxuriant gardens.[4]

Adding to Longfellow's flowery verse—capturing the river's charm, unpredictability, and rich habitat—were explorers and settlers, such as Timothy Flint, who reiterated the Mississippi's place in the expansive nation. In Flint's words, "No person, who descends this river for the first time, receives clear and Adequate ideas of its grandeur, and the amount of water which it carries." Journalists joined the chorus of praise as T. B. Thorpe articulated the sense of exceptionalism and the river's promise for the future. In his words,

> The vast country it drains, the rapid influx of population into its fertile valleys, the wonderful enterprises of the people, the development of wealth, the triumphs of steam, the progress of empire, have no precedents in the past, and there can be nothing equal to it in the future.[5]

Adding to the written litany of praise was the work of popular artists such as George Caleb Bingham, whose Mississippi River landscapes promised a place in the American imagination as images of flatboatmen became synonymous with the unconventional, a "rough and ready" American frontier. Paralleling Bingham's imagery were mid-nineteenth-century panorama portraits by artists such as John Banvard. Thousands waited in line, in cities as far away as London, to view his *Panorama of the Mississippi*, an artwork extending for 3 miles. With the arrival of steam, another Mississippi River emerged, dominated by steamboats carrying goods and people to ports such as New Orleans, St. Louis, Cincinnati, Memphis, and Natchez. In the steamboat era, the imagery not only portrayed the nation's major artery but also reinforced popular beliefs about technology and progress. The demise of steam, by the late nineteenth century, only to be replaced by rail, coincided with a river that needed to be mastered. Devastating floods, such as those that occurred in the late nineteenth century and early twentieth century, particularly the 1927 Flood, confirmed the river's independence, provoking an outcry to subdue the river. A new era of levee building began, and the engineered Mississippi River was born: an icon of technological expertise.[6]

In all these mediums—history, poetry, and art—several tropes endured. First, the river was perceived as an empire-builder, the highway that connected the vast American interior. Second, the river was often framed as a nurturer, cultivating a rich, garden landscape. Third, in an abstract sense, the river served as a mediator between the natural world and civilization. In its role as empire-builder, the river was recognized and portrayed as powerful, forbidding, and unreliable. Yet, the river as nurturer watered "luxuriant gardens." Further, in the southern reaches of the river basin, another correlation with the river emerged as the surrounding bayous and swamps were to the Europeans and later American settlers, a no-man's land, the boundary between a tamed and untamed landscape. The distinctions, often in direct conflict, persist. For example, in recent years, the river and its many tributaries, such as the Ohio, have served as tourist attractions, with river cities such as Cincinnati, Ohio investing in riverfront parks. In the case of Cincinnati, located on the banks of the Ohio River, the redesigned riverfront includes a history of prominent steamboats that graced its shores. Yet this same riverbank served as a public landing where enslaved people were transported. The first is commemorated, whereas the second is omitted.

As the contemporary case of Cincinnati illustrates, despite widespread sentiments about the river and its environs, they were not shared by all people living and working along the Mississippi River. Beginning with the arrival of enslaved Africans in the early 1700s, another river narrative evolved over the next several centuries. At times, their perceptions of the river converged with those of the colonists, and later white settlers, in that all acknowledged the power of the river as it jumped riverbanks, unleashed devastating floods, and challenged all who tried to navigate it. However, the Euro-American inclination was to develop and master the river, usually in the name of commerce, whereas the enslaved African American experienced the river as the site of escape. Many of the first

Africans brought to the lower Mississippi River Basin found within the dense Delta lands—the bayous adjacent to the river—a means to escape from slavery as numerous accounts survived commenting on runaways, called maroons, and their flight to the surrounding bayous. Within the slave codes in the Mississippi River Basin—beginning with the French code noir and continuing with each successive empire—the treatment for runaways was always included, indicating the frequency of these bids for freedom. The experience of the river as a means of escape continued throughout the colonial and antebellum period as slave narratives chronicled their time evading capture in the swamplands.[7]

Another example of contrasting narratives can be found in the nineteenth and twentieth centuries' imagery of levees. To engineers, white Southern planters, and community boosters, the construction of levees prevented flooding and facilitated commerce, thereby protecting cotton production and other goods, shipped on the Mississippi River. As the engineering community matured, levees also represented modernization; progress was derived from technological expertise. But to African Americans in the antebellum period, levee construction signified hard, back-breaking labor for men and women. By the late nineteenth century, the hypocrisy of "progress" was further questioned as the levees for black Americans were symbols of rough, raucous work camps where life was short and valued little. The differences in perspectives persisted throughout the twentieth century, as the 1927 Flood exposed a Jim Crow South where African Americans were relegated to sharecropping marginal flood-prone land and forced to remain in the flooded areas where lives were sacrificed to save levees. But the 1927 Flood also revealed levees as sites of refuge for both African Americans and whites.[8]

Thus, in chronicling the history of the Mississippi River through the experiences of African Americans who lived, worked, recreated, revered, and despised the river, a river history distinct from earlier Mississippi River histories emerged. Throughout the following chronological narrative, for each era—the colonial, antebellum, post–Civil War, early twentieth century, and modern era—this text benefited from new and exciting scholarship from multiple disciplinary perspectives. For Chapter 1, the narrative begins as Africans are forcibly brought to New Orleans, beginning two centuries of enslaved labor digging ditches, navigating the river, and undertaking other jobs to sustain the French presence in an unreceptive environment. As the empire's commitment to the Lower Mississippi Valley grew, the French enslaved large numbers of Africans to grow rice, tobacco, and indigo. Although enslaved, many found means to supplement a meager diet as they became familiar, often as river pilots, with the riverine environment and would harvest the nearby cypress swamps for logs and Spanish moss. But life in the riverine environment also offered opportunities for escape as marronage occurred frequently during the French colonial period. Considered uninhabitable by the colonists and later, southern planters, the swamps offered another space in the story of African Americans' relationship with the river. According to Gwendolyn M. Hall, one of the preeminent scholars of colonial Louisiana, the "lower-river settlements had become a haven for maroon raiders: fugitives from estates ranging from Point Coupee to below New Orleans." Often, whole families disappeared into the swamps, with some finding sustenance by harvesting cypress and supplying nearby markets. Because of the forbidding nature of the terrain, maroon communities often existed for years, as Hall described in her seminal work, *Africans in Colonial Louisiana*. Using court records and secondary sources, such as Hall's work and more recent additions, such as Sylvia Diouf's text, *Slavery's Exiles*, a history of marronage is explored in Chapters 1 and 2.

Yet when citing secondary sources on marronage, most scholars have relayed this history through the lens of collective resistance and of course, within the broader history of slavery. For example, Hall emphasized the geographical origins of newly arrived Africans. Her research illustrated the importance of African geography and culture and their subsequent effect upon the African experience in colonial Louisiana. In the following work, however, the maroon communities and the individual maroons offered another perspective in the story of African Americans' relationship with the Mississippi River. Within this space, there are real and psychological dimensions: real in the physical presence of the swamplands and bayous of the Mississippi River Valley and psychological in the sense of refuge that these lands symbolized for enslaved people in succeeding generations, as slaves lived, raised children, and created economies that relied upon skills in extracting resources from uninviting cypress swamps or woodlands. The knowledge required to live in these demanding environments is admirable but to secure a living for a number of years, even more so. How the maroons negotiated the spaces that surrounded them became legendary, fueled by heroic figures, such as San Malo. In the case of San Malo, through his skills and tenacity, he succeeded in leading a maroon community for more than a decade, until his capture and subsequent execution by Spanish authorities. Thus, San Malo and his maroon community illustrated the themes that would characterize African Americans' relationship with the Mississippi River and its surroundings. For San Malo—whose exploits were memorialized in the "Ballad of San Malo"—the river and its swamps represented freedom, cultivating a resilience in the African American community that persisted for the next two centuries.[9]

In Chapter 2, beginning with the Louisiana Purchase and the U.S. presence in the Mississippi River Valley, the institution of slavery becomes entrenched along the river from the state of Missouri to the Gulf. Although free blacks lived and worked in the region, the majority of African Americans were enslaved. Yet despite bondage, the antebellum period uncovered a history where power through the appropriation of space was always being mediated. Once again, the bayous and swamplands were drafted into service for clandestine meetings, or sites of refuge, whether temporary or permanent, allowing enslaved people to cultivate an autonomy through the mastery of lands surrounding the plantations. Included throughout will be the studies of other scholars that recognize the existence of alternate geographies in the enslaved person's world and thus, the connections between identity and the environment. To borrow a term from Edward Said, these "rival geographies" allowed for resistance in the enslaved people's community or offered a space for amusement. Stephanie Camp, in *Closer to Freedom*, recognized Said's contributions when examining the enslaved person's existence. She detailed how enslaved people developed "alternative ways of knowing and using the plantation and southern space." Rashauna Johnson, in *Slavery's Metropolis* relied upon Camp's work in her study of enslaved people during the late eighteenth and early nineteenth centuries. But there are others who acknowledged the presence of differing geographies, such as Melvin Dixon in *Ride Out the Wilderness*, who observed: "During slavery blacks depicted the wilderness as a place of refuge beyond the restricted world of the plantation." While their works converged with this text in the recognition of these spaces or alternate geographies, the focus here is different. For the purposes of the following text, the space itself has significance as it contributed to a competing narrative of the river. The "rival geographies" of enslaved people produced another set of relationships or linkages with the river, culminating in a history far different than that of a steamboat captain or planter.[10]

Yet the river, itself, evolved as a symbol of oppression carrying enslaved people into the bowels of one of the most cruel slave regimes. In contrast, however, for a robust, capitalist economy, the slave regime of the Deep South was wildly prosperous through the production of cotton and sugarcane with the river facilitating. The same river, however, that oppressed also represented freedom as an escape route or as Thomas Buchanan demonstrated, through laboring on the river. Although many working on steamboats or laboring as roustabouts were still enslaved, they enjoyed a freedom uncommon to those in bondage on the plantations. For studies that specifically explore the relationship of black laborers and the Mississippi River, two works considered were those by Thomas C. Buchanan and Daniel Usner. Buchanan's *Black Life on the Mississippi* examines African American life during the time of slavery, where he considered the world of steamboat travel. By looking at the lives of blacks working on the steamboats, he found that in comparison to other types of labor, this work often had a liberating effect. On the steamboats, blacks were not under the ever-watchful eye of the overseer and had more opportunities for escape. Also, for those who worked on the boats, they served as a conduit for other African Americans living up and down the river as they provided a network, relaying information from one group to another. In Usner's works, in which there are several, he reveals the diversity of jobs that African Americans (enslaved and free) performed. In the Mississippi Basin, blacks might be dockworkers, boatmen, agricultural laborers, artisans in New Orleans, and firemen on steamboats, to name a few. Usner's research offers further proof of how labor informed the collective memory of African Americans living along the Mississippi River.[11]

The narratives of self-emancipated African Americans also illuminated the river as escape, while highlighting their resilience as they journeyed north. Part of their resilience derived from a spiritual perspective as following the North Star or crossing "over the River Jordan," held symbolic meanings for African Americans. Slave narratives offered fascinating records of escaping through the swamps or securing passage on a northbound steamship. In both instances, whether recounting the history of maroons or runaway slaves whose journey included crossing swamps or dense woodlands, advertisements posted by enslavers also furnished insights into this history.[12] But the maroons that lived in the lower Mississippi River Valley during the antebellum period revealed another aspect of the intersection of African Americans and the river. In order to live in the borderland or hinterland with its overgrown vegetation, an exceptional skill set is required. Beginning with San Malo, how the maroons negotiated these spaces is another glimpse into competing narratives of the Mississippi River. But in recognizing the necessary skills to navigate the challenging landscapes, one caveat remained in that many formerly enslaved people realized when recounting their escape, the skill level had to be deemphasized. For if the formally enslaved person touted his abilities in the forbidding swamps, he lent credence to a nineteenth-century stereotype that blacks were uncivilized and primitive.[13]

Yet despite the mastery of the maroons, the autonomy gained through the appropriation of space, or the steamboat's opportunity for freedom, the river remained a site for violence and trauma as Cynthia Gooch's research demonstrated. The bloody aftermath to San Malo's attempts to evade Spanish authorities would be revisited many times over in the antebellum period. For the enslaved African Americans who revolted, such as the those in the little-known Deslondes Revolt, their capture ended with severed heads perched upon pikes on the levees lining the lower Mississippi River. Others, unsuccessful in their bids for freedom, also met with brutal, inhumane punishments and often, death. Still despite having been enslaved, self-emancipated William Bibb could view the river and appreciate its aesthetic value, while questioning why the freedom allowed to nonhuman species was not allowed for all.[14]

In concluding the antebellum period, Chapter 3 explored the post–Civil War world. Again, the Mississippi River occupied a significant place in the lives of many African Americans revisiting the common intersections of the river as liberator and oppressor. For the Exodusters, a group of African Americans intent on leaving Mississippi, Louisiana, and some Midwestern states in the late 1870s, the river offered passage to "free land" in Kansas while the levees provided refuge along the way. Their stories of leaving the post–Reconstruction South, first captured by Nell Irvin Painter in her work, *Exodusters*, illuminated a new era where white Southerners sought to regain privilege and power. Other scholars, such as Bryan M. Jack, have expanded Painter's research, with his study focusing on the Exodusters and the African American community in St. Louis. The commitment of the Exodusters to pursue new lives in uncharted territory testified to the ongoing violence of the South, as new practices such as "bulldozing" became a common means to disenfranchise African Americans. As a result, African American lives were constantly under threat, politically, socially, and economically.[15] This chapter also examined white Southerners' efforts, in a world where slavery was abolished, to devise new ways to extract labor at little to no expense. One of the most pressing areas in need of repair was the river's levee system, particularly after so many were destroyed during the Civil War. In response, levee work camps were established. The labor for the camps often came from a revamped privately leased prison system where thousands of African Americans were incarcerated for minor offenses. Known for their brutality, the camps were little better than the earlier slave regime. Scholars such as Mark T. Carleton and David M. Oshinsky have documented the abuses in Louisiana and Mississippi, whereas more recent works from scholars such as Tabitha LaFlouria present a prison system determined to establish a post-war white hierarchy. Adding to their scholarship are blues songs such as "Levee Camp Moan," which memorialized the hardships endured.

However, the post-war world also meant laboring on the steamboats that still populated the river. Tales of roustabouts' exploits and unconventional lifestyles characterized the era accompanied by accounts of impoverishment and dire work conditions. Through song, newspaper accounts, and oral histories, the steamboat workers' lives are recounted. Throughout these decades, the consistency of the river's presence is evident. For African Americans, their associations with the river, conveyed through song, prose, and folklore, remained the same. The river represented freedom or rebirth as seen through the Exodusters; oppression and violence as experienced by those in the brutal levee camps, the demanding work of the roustabout, or the sharecropper forced to accept unrealistic terms and a river prone to flooding.[16]

By the time of the 1927 Flood, the subject of Chapter 4, many African Americans living along the river were consigned to marginal lands and would be disproportionately affected by the flood. The 1927 Flood, unrivaled in scale from previous floods, displaced 700,000 people. But in addition to losing their homes and belongings, African Americans met with harsh treatment at the relief sites and often were coerced into the futile work of saving the levees. Other injustices included being forced to remain in the area as white Southerners worried about retaining their workforce after impressing black labor to save the levees. The unequal treatment extended to areas designated as shelters—makeshift lodgings—as white flood victims were often housed in more comfortable surroundings.[17] These disparities informed the cultural production of song as a blues genre known as flood blues lamented the treatment of African American flood victims while recognizing the river's agency in the destruction of lives and homes. In studying this genre, Alan Lomax's research, although not without issues, is a starting point. In his *The Land Where the Blues Began*, Lomax spoke with numerous blues musicians and recorded their songs for posterity. His book is an encyclopedia of the flood blues along with songs depicting work on the levees. Following Lomax and applying more of a scholarly approach to understanding the blues in which the "flood blues" emerged is Paul Oliver's text, *The Blues Fell this Morning*. In Oliver's review of the blues and their popularity in the early twentieth century, he documented the sales of blues records through the record industry's production of what were called "race records." His work demonstrated the popularity of this musical genre and as a result, the role of the Mississippi

River in this discourse. More recent literature can be found in the works of well-known music experts such as David Evans' *Big Road Blues: Tradition and Creativity in the Folk Blues* and Robert Springer, who edited, *Nobody Knows Where the Blues Come From: Lyrics and History*. But for a historical perspective of the blues' importance as revealed through the 1927 Flood, Robert Mizelle, Jr.'s text *Backwater Blues* is the most informative.[18]

For many African Americans, the 1927 Flood—similar to Katrina in later years—represented a continuity in their relationship with the river. Once again, the river represented escape, this time to start over as seen in the migration north. In addition to music, literary works framed collective memories. For example, in Richard Wright's classic short story, "The Man Who Saw the Flood" (1940), Wright's description of the 1927 Flood was moving as he chronicled the unrelenting poverty associated with life along the Mississippi, only to be exacerbated by floods. In Wright's work, he encapsulated the experiences of the African American community as the 1927 Flood became another historical marker in a history of lives interacting with the river. In some instances, African Americans saw the flood as symbolic—washing the slate clean and affording the opportunity to leave and start anew. Others, however, resigned themselves to lives of tenancy and sharecropping. Complementing Wright's work are the works of other novelists that provided insights into the cultural space the river occupied for African America. Yet the 1927 Flood revealed a familiar river for African Americans, whereas for white residents who lived along the riverbanks, the flood further solidified the need to manipulate and control the nation's waterway. The divergent reactions generated differing narratives, shaping distinct cultural memories. For African Americans, the flood blues and short stories, such as Wright's, framed a river that could be cold, unyielding, and prompting flight for many. Later floods reiterated this imagery, as the 1930s' news accounts revealed.[19]

Although river-related disasters persisted throughout the twentieth century, African Americans sustained their communities along the Mississippi River. As Chapter 5 outlined, communities such as Fazendeville in the Lower Mississippi Valley, occupying land that once saw the Battle of New Orleans, enjoyed the largesse of the river for 100 years. Fazendeville residents recalled fishing in the river and growing gardens that allowed for a self-sufficiency for each successive generation, whereas other memories included walking down to the river for baptisms, overseen by Battle Ground Baptist Church. These communal traditions, however, were halted when Fazendeville was razed, privileging the memory of a battlefield over that of a long-standing community. In response, many residents moved to the nearby Ninth Ward in New Orleans, another primarily African American community, located by the river. The Ninth Ward residents, sometimes representing several generations, recall an environment where they built homes from cypress logs, harvested from the surrounding swampland, fished from the river, and worked in river-related jobs. In the words of one resident, the community they built was "a little world unto itself." But the effects of Hurricane Betsy in the 1960s, followed by the devastating Hurricane Katrina and its life-threatening storm surge, forced many to leave. In the case of Betsy, the displacement was temporary but with Katrina, the move away for some has been permanent.[20]

Not surprisingly, scholars have drawn parallels between Katrina and the 1927 Flood when the losses for African Americans were disproportionately greater. Flashbacks to 1927 ranged from the federal government's indifference to the plight of so many African Americans to the government's preferential treatment to white flood victims to the perception that certain levees were blown up to save white neighborhoods and destroy black communities. Unlike 1927, however, the documentation of the abuses was more extensive and not limited to African American publications or whitewashed Red Cross accounts. Instead, the literature on Katrina is profuse with commentaries from historians and journalists such as Douglas Brinkley and Michael Dyson, both offering insights soon after the disaster to the more recent assessments by Andy Horowitz written more than 15 years later. But there were other similarities—Katrina prompted an outpouring of responses from artists, musicians, poets, and writers. Through social media outlets, professionals and nonprofessionals alike reacted to the images of floating dead bodies, displaced families, and the ill-equipped disaster relief site at the Superdome. Yet, despite the scale of suffering after Katrina, the desire to return to neighborhoods, such as the Ninth Ward, and rebuild was strong. In numerous interviews, residents spoke of their wish to move back, often to homes that had been in the family for multiple generations. There were, however, other survivors, who, like those in 1927, chose to remain in their new communities, forging new lives in cities such as Houston and Atlanta. But the parallels do not end here as Katrina revealed another recurring theme. Sites such as the Ninth Ward experienced greater damage because of its geography as residents occupied land that was once wetlands, and that was thus more vulnerable to flooding. The Ninth Ward was also one of the last areas to be rebuilt. Despite the setbacks, for those who returned, the success stories—drawn upon centuries of perseverance—are abundant. For example, the Battle Ground Baptist Church at Fazendeville, a long-time fixture in the Ninth Ward, would be one of the organizations committed to rebuilding as it had done so many times in the past. Numerous other firsthand accounts testified to the determination of so many to return and rebuild, a resilience that has characterized African Americans' relationship to the river and its environs.[21]

Unfortunately, devastating hurricanes such as Betsy, Katrina, and later Ida were not the only threat to African American communities living along the river. As the Epilogue detailed, by the 1980s, many riverfront communities from Baton Rouge to New Orleans were confronted with the presence of petrochemical giants. This explosion-prone industry generated toxic emissions, causing innumerable health problems for those living nearby and earning the river corridor the sobriquet, "cancer alley." The African American communities of Old Diamond, Geismar, and Mossville, to name a few, experienced higher than average rates of cancer, kidney and respiratory problems, and birth defects as a result of the giant industries' proximity. Some of the communities had existed since the Civil War when many African Americans chose to remain on former plantation lands. For those who stayed, their ancestors were buried on the properties and as they grew up in the communities, they fondly recalled childhoods where the levees and swamps were an integral part of their landscapes. In response to the presence of petrochemical companies such as Royal Dutch Shell and Du Pont, residents formed grassroots organizations with goals ranging from the prevention of any further industry development to accountability on the part of the industrial behemoths. Assuming responsibility often translated into a reduction of emissions or relocation costs for residents. These local organizations, building upon a civil rights tradition of resistance, included groups such as Concerned Citizens of NORCO, Concerned Citizens of St. John Parish, and RISE St. James. Savvy in their strategies, with local leadership by internationally recognized environmental activists such as Sharon Lavigne and Margie Richard, for example, they attracted national and international attention as their circumstances reflected the environmental racism that scholars such as Robert Bullard and Rob Nixon have identified. By 2021, the influential United Nations was calling for an end to environmental racism in "Cancer Alley."

Accompanying the threat to individual communities, the Lower Mississippi Valley entered the twenty-first century, another casualty of the Anthropocene Age. The alarming loss of wetlands, due to human decisions, threatened coastal communities as well as increased the damage sustained by hurricanes. As canals wind through swamplands, houses are built on former wetlands, and a system of locks and dams in the upper stretches have claimed the landscape, a different Mississippi River has emerged. This Mississippi River and its environs would be unrecognizable to the river's first inhabitants. Thus, in the present era, while African Americans' history with the Mississippi River and its surroundings reflects certain constants, new health and environmental challenges usher in an era with many questions yet to be resolved. The Mississippi River Valley, similar to so many other vulnerable areas across the globe, is under siege in ways unseen by previous generations.[22]

As the chapter overviews indicate, this study is first and foremost a narrative history of the Mississippi River, through the lens of African Americans who knew and interacted with a river that threatened and enriched lives. In addition to the content expertise of scholars versed in Mississippi River history, other subdisciplines have been consulted in writing this book. Although not an environmental history, the contributions of the subdiscipline have informed and supported the research throughout. Beginning with Carolyn Merchant's call for the inclusion of race when formulating environmental history, other scholars have followed. But Merchant was one of the first to recognize that "American Indians and African Americans perceived wilderness in ways that differed markedly from those of white Americans." She also went on to draw the parallels between the oppression of enslaved African Americans and the degradation of the land under cotton, tobacco, and rice production. Adding to Merchant's conclusion is the work of Carolyn Finney. In Finney's provocative work, she explored America's dominant environmental or wilderness narrative and how it has excluded African American connections. The crafting of the narrative is important as it informs national identity and by omitting African Americans, as in not only are they absent in the narrative of racialized space but their experiences with the natural world have also been lessened and demeaned. To redress the narrative, Finley, an African American scholar, advocated the inclusion of individual and collective memories to "recreate and form our own environmental spaces." The following Mississippi River history, drawn from the perspectives of African Americans, can contribute to an integrated history where African American experiences with the natural environment become as familiar as Old South mythologies. In other environmental history accounts, Mart Stewart and Kimberly Smith articulated the differing perspectives between African Americans and white Americans. Comparing the environmental horizons of enslaved people, Stewart successfully argued how not only did the enslaved know their physical surroundings better than the planters, but also this knowledge was empowering. Although Stewart was not referencing the Mississippi River landscapes, circumstances would be similar, as argued in Chapter 2. Kimberly Smith furthered these arguments as her text on black environmental thought considered how African Americans viewed the wilderness. In her conclusions, for enslaved people, the wilderness would never embody a pristine geography but instead remained the negotiated space between the enslaved and enslavers.[23]

Still other notable works in environmental history and the humanities include the only reader in African American environmental history. Dianne Glave and Mark Stoll edited a volume that has been very useful for this study, with articles ranging from the turpentine industry to contemporary discussions of environmental racism. For example, one of the articles by Scott Giltner on hunting and fishing practices during the antebellum period provided new insights on the knowledge enslaved people possessed about their surroundings and the practical nature of hunting to supplement diet. In considering how African Americans perceived their environment through text, Camille T. Dungy assembled several centuries of African American nature poetry. She included a poem by George Marion McClellan, written in 1916 and titled "A September Night." The setting is Antigua, Mississippi and the following lines capture the beauty of the river and its bayous:

> The full September moon sheds floods of light,
> And all the bayou's face is gemmed with stars,
> Save where are dropped fantastic shadows down
> From sycamores and moss-hung cypress trees.
> With slumberous sound the waters half asleep
> Creep on and on their way, 'twixt rankish reeds,
> Through marsh and lowlands stretching to the Gulf.

Through Dungy's overview of African American poetry, she reminds the reader, "Even during the most difficult periods of African American history, the natural world held potential to be a source of refuge, sustenance, and uncompromised beauty." Through her work and that of other environmental historians, our understanding is furthered that the intersection of African American lives with the river and its surroundings was never one-dimensional.[24]

Adding to considerations regarding the role of race in environmental history and strengthening the text was recent scholarship on the intersections of labor and environmental history. Richard White's essay "Are You an Environmentalist or Do You Work for a Living?" sets the stage by encouraging environmental historians to "reexamine the connections between work and nature." In White's classic, *The Organic Machine*, which traces the history of the Columbia River, he brought the Columbia River alive as memories of the river were retrieved and integrated into the river that exists today, departing from previous environmental histories that emphasized decline and loss. White looks at how energy was the linkage, the connector between humans and the river. In his history of the first explorers to the Columbia, trying to navigate upstream, he observes how they experience the river through the energy they expended. This would have been a familiar experience for the first African Americans who served as river pilots on the Lower Mississippi River. Since White's work, there have been others, including Gunther Peck's "The Nature of Fault Lines and Common Ground in Environmental and Labor History," an analytical work in *The Journal of Environmental History* that finds commonalities between labor and environmental history. Still another work is Chad Montrie's *Making a Living: Work and Environment in the U.S.*, an anthology of scholars examining the linkages between labor and the environment in the United States from the times of slavery to the twentieth century. However, in each of his selections, the workers "confronted an industrial transition," whereas Stefania Barca's recent article, "Laboring the Earth: Transnational Reflections on the Environmental History of Work" in *The Journal of Environmental History* identifies three junctures where work and nature intersect and offer environmental history "possible new paths of investigation." Although useful, these frequently referenced works are focused on the arrival of industrial capitalism and how its wage labor class alienated workers from nature (Barca's work is the exception.).[25]

Closer to the experiences of African Americans is Joanna Dyl's work, "Transience, Labor, and Nature: Itinerant Workers in the American West," in *International Labor and Working Class History*. Dyl considers the lives of hoboes and while she looks at them within the context of industrial capitalism, she goes beyond this frame as she also examines their appreciation of wilderness, thereby challenging Roderick Nash's critique that "lumberman, miners, and professional hunters . . . lived too close to nature to appreciate it for other than its economic value as raw material." This experience corresponded with other marginalized groups living and working in a riverine environment, lending support to slave narratives such as Bibb's and his aesthetic appreciation of the river and the nonhuman species it supported. Dyl's work is also of comparative value for African Americans' environmental perspective, in that similar to the hoboes and other marginalized groups, African Americans "viscerally experienced outdoor living and the vagaries of climate." But the work that is the most applicable to the lives of African Americans laboring on the river and its surroundings is Thomas G. Andrew's work, *Killing for Coal*, in which he traces the history of the 1914 Ludlow Massacre and the "Great Coalfield War" with an in-depth look at the lives of the colliers. By studying this group of laborers and the work they performed underground, Andrews offers a new framework for environmental historians. He coined the term "workscape" as opposed to landscape as in his words,

> something more complex: not just an essentially static scene or setting neatly contained within borders, but a constellation of unruly and ever-unfolding relationships—not simply land, but also air and water, bodies and organisms, as well as the language people use to understand the world, and the lens of culture through which they make sense of and act on their surroundings.

By looking at the colliers' experiences through workscapes, Andrews blurs the artificial distinction between humans and nature and instead shows how each influenced the other; this was certainly the case for African Americans laboring on the river. Andrews' text also goes beyond a common theme of placing preindustrial labor's relationship with the environment as one that emphasizes sustainability or at best traditional ecological knowledge. But whether in the context of industrial capitalism or Andrew's workscapes, all these environmental historians agree that the relationship between labor and the environment warrants study, a relationship that dominated African Americans' experience with the Mississippi River.[26]

Another relatively recent field of study that furthers our understanding of labor and nature is the works of scholars studying capitalism in the United States during times of slavery. These provocative scholars, such as Walter Johnson and his work *River of Darkness*, exposed a robust American economy in cotton and sugar commodities grown in the Mississippi Delta and how enslaved labor facilitated these lucrative enterprises. Other scholars successfully reframing the story of slavery within the context of capitalism include Sven Beckert and his award-winning *Empire of Cotton* and Edward Baptist's *The Half Has Never Been Told*. Through their studies, demonstrating the immense profits realized by Southern planters, the parallels between the harnessing of the river and the degradation of the environment correspond heavily with the oppression of a labor force, in this instance enslaved labor. Paul Outka's scholarship, in *Race and Nature*, reinforced this link as he argued that the degradation of enslaved Africans coincided with the despoliation of the environment. Both white planters and capitalists perceived nature—which to them included the enslaved Africans—as an exploitative, expendable commodity.[27]

Another recent work that explored this connection is Thomas D. Rogers' *The Deepest Wounds*. In his study of the sugarcane industry in Pernambuco, Rogers considered the planters' environmental perspective and contended that when looking at their estates, they saw "laboring landscapes," which encompasses all facets—the land and workers. For the laborers, growing and harvesting the cane, the "crucial metaphor was captivity, and it hinged on their long experience of exploitation on plantations." Rogers enlisted an array of sources, including oral histories and folklore, to expose the workers' perspectives. Finally, by looking at both groups—workers and planters—he revealed a landscape where a growing sugarcane industry exploited the land and workers while also illustrating the differing perspectives when viewing their landscapes. His text is significant for demonstrating the importance of wedding the histories of all groups living and working within a specific landscape. Rogers' work further underscored the need for an African American history of the Mississippi River, thereby integrating the histories of all who experienced this riverine environment. Finally, Rogers illustrated the comparative potential of studies examining the linkages between labor and the environment.[28]

Ultimately, aided by contributions from these scholarly works, this is a text about the Mississippi River through the lens of African Americans' experiences with the river. These experiences include years of bondage, laboring on sugarcane and cotton plantations, steamships, levees, and cypress forests in Mississippi bayous while experiencing disproportionate consequences from an ecological regime with frequent flooding, not to mention the occasional hurricane. These experiences, whether before or after the Civil War, were linked to the river, producing a river that looks a little different than the one preserved in Southern and frontier mythology, whether it be the swashbuckling raftsman or the genteel world of Southern plantation owners. Further, there is a constancy—a trajectory—to the history as parallels can be drawn between those displaced after Katrina and those displaced by the 1927 Great Mississippi Flood. But unlike other works looking at the African American experience, in this text the Mississippi River is also the focus and how the river and its surrounding landscape were perceived are part of the story. Questions such as "What did a swamp, a bayou mean to other groups?" will be posed. To African Americans, the swamp represented escape, freedom, a space outside the managed landscape of the planter, for example. Other considerations include, how did the river represent an unfiltered means of communication for African American workers and the black community living along the river? What did levees represent? To those African Americans that labored on them, levees represented unceasing work and in times of flooding, further constraint. For African Americans who lived in riverfront communities, levees might represent refuge and protection. By the twentieth century, levees also translated into a loss of land. However, to the Southern planter or politician, the levee represented the ever-sought-after internal improvements, funded by federal largesse. At certain junctures, however, the perceptions of the Mississippi converged for those living and working along the river. For example, understanding the rhythms of the river was a source of pride for white and black boatmen. The grandeur of the continent's longest river was appreciated by all as was the sight of the first steamships, to name a few shared perceptions. But more often than not, the experiences recounted in this text were unique to African Americans, revealing another river and its history.

While this work will cite the expertise of well-known historians versed in U.S. history, from the colonial to the modern period, there will be omissions as the emphasis is on reframing how the river is viewed. How do we accurately portray the Mississippi River? My hope is that in future works, we see the river in its entirety so when discussing the rich, palatial steamships plying the river, images of African Americans feeding the incessantly hungry boilers or roustabouts unloading the boats also figure prominently. In addition to a fuller history of the Mississippi, another insight might be drawn from understanding the river from the perspective of those whose lives were marginalized by the dominant culture for those groups often interacting more closely with the river, affording us glimpses of a relationship with nature that will be critical to understand today. For by living in the swamplands as maroons or learning the rhythms of a river as boat workers or deriving a living from the land, a knowledge of this river system emerges that is absent from the narratives of Southern planters, northern engineers, tradesmen, not to mention the poets, artists, and writers. The following chapters represent an effort to address the gaps and fill out the story of the Mississippi River by including the lives of others who transformed and were transformed by the riverine landscape.

Notes

1 Major works on the Mississippi River include the environmental histories by Christopher Morris, *The Big Muddy: An Environmental History of the Mississippi and Its Peoples from Hernando De Soto to Hurricane Katrina* (New York: Oxford University Press, 2012); Mikko Saikku, *This Delta, This Land: An Environmental History of the Yazoo-Mississippi Floodplain* (Athens: University of Georgia Press, 2005). For a cultural history approach to the river, consult Thomas Ruys Smith's publications, beginning with *River of Dreams: Imagining the Mississippi before Mark Twain* (Baton Rouge: Louisiana State Press, 2007). Popular histories include Paul Schneider, *Old Man River: The Mississippi in North American History* (New York: Henry Holt and Company, 2013); Lee Sandlin, *Wicked River: The Mississippi When It Last Ran Wild* (New York: Vintage Books, 2010). For a well-written memoir chronicling an African American's journey down the river, see Eddy L. Harris, *Mississippi Solo: A Memoir* (New York: Henry Holt and Company, 1988).

2 Multiple texts examine the role of rivers in shaping cultures and civilizations. Two notable ones include Tricia Cusack, *Riverscapes and National Identity* (Syracuse: Syracuse University Press, 2010); Peter Coates, *A Story of Six Rivers: History, Culture, and Ecology* (London: Reaktion Books, 2013). For individual river texts, scholarly works include Elaine Sciolino, *The Seine: The River That Made Paris* (New York: W.W. Norton and Company, 2019); Ruth Mostern, *The Yellow River: A Natural and Unnatural History* (New Haven: Yale University Press, 2021); Sara Pritchard, *Confluence: The Nature of Technology and the Remaking of the Rhone* (Cambridge: Harvard University Press, 2011); Marc Cioc, *The Rhine: An Eco-biography, 1815–2000* (Seattle: University of Washington Press, 2002).

3 Another river that enjoyed celebrity status in the United States was the Hudson River, prompting the Hudson River School (1815–1876) with artists such as Thomas Cole and Frederic Edwin Church portraying the scenic Hudson River Valley. For further discussion, see New York Historical Society, *Hudson River School: Nature and the American Vision* (New York: Rizzoli Electa, 2009); Dean Klinkenberg, "The 70 Million-Year-Old History of the Mississippi River" *Smithsonian Magazine* (September 2020) accessed at https://www.smithsonianmag.com/science-nature/geological-history-mississippi-river-180975509/; *The De Soto Chronicles, the Expedition of Hernando de Soto to North American in 1539–1543*, vol. 1, ed. Lawrence Al Clayton, et al. (Tuscaloosa: University of Alabama Press, 1993):113.

4 For a discussion of Mississippi River imagery in the Early National Period, see John Seelye, *Prophetic Waters: The River in Early American Life and Literature* (New York: Oxford University Press, 1977). Henry Wadsworth Longfellow, *Henry Wadsworth Longfellow: Poems and Other Writings*, Library of America Series (New York: Library of America, 2000).

5 Timothy Flint, *The History and Geography of the Mississippi Valley*, Vol. 1, 3rd ed. (Boston: Carter, Hendee and Co., 1833):93–94; Thomas Bangs Thorpe, "Remembrances of the Mississippi" *Harpers New Monthly Magazine*, Vol. 12 (December 1855–May 1856):27.

6 John Francis McDermott, *George Caleb Bingham: River Portraitist* (Norman: University of Oklahoma Press, 1959). To learn more about Banvard's panorama and other Mississippi River representations, see Phillip Gentile, "Viewing the Iconic Mississippi: Strategies of Reenactment in River Panoramas and Bill Morrison's *The Great Flood* (2013)" *Southern Quarterly*, Vol. 53, No. 3 (Spring 2015):121–136; Janet L. Whitmore, "A Panorama of Unequaled Yet Ever-Varying Beauty" *Currents of Change: Art and Life Along the Mississippi River, 1850–1861*, eds. Jason Busch and Christopher Monkhouse (Minneapolis: Minneapolis Institute of the Arts, 2004):12–62; Bernard Comment, *The Painted Panorama* (New York: Harry N. Abrams, Inc., Publishers, 2000).

7 T. Wilson, 1858 Interview in *Slave Testimony: Two Centuries of Letters, Speeches, Interviews and Autobiographies*, ed. J.W. Blassingame (Baton Rouge: Louisiana State University Press, 1977):340.

8 While the origins of this book were in place before the birth of the 1619 Project in 2019, the following text complements certain facets of the 1619 Project in that by looking at the Mississippi River through the lens of Black Americans, the hypocrisy of "progress" is revealed. For example, in the nineteenth century, for groups such as the engineering community, white Southern planters, and community boosters, the construction of levees meant flood prevention, thereby protecting the business of cotton production along the river. But to African Americans, levee construction signified hard, back-breaking labor for men and women. Differences such as this persisted throughout the next century. See Nikole Hannah-Jones, "The 1619 Project" *New York Times Magazine*, 14 August 2019.

9 Gwendolyn Midlo Hall, *Africans in Colonial Louisiana: The Development of Afro-Creole Culture in the Eighteenth Century* (Baton Rouge: Louisiana State University Press, 1992):9, 115; Sylvia Diouf, *Slavery's Exiles: The Story of the American Maroons* (New York: New York University Press, 2014).

10 Edward Said, *Culture and Imperialism* (London: Chatto & Windus, 1993); Stephanie Camp, *Closer to Freedom: Enslaved Women and Everyday Resistance in the Plantation South* (Chapel Hill: University of North Carolina Press, 2004):3, 7; Rashauna Johnson, *Slavery's Metropolis: Unfree Labor in New Orleans During the Age of Revolution* (Cambridge: Cambridge University Press, 2016); Melvin Dixon, *Ride Out the Wilderness: Geography and Identity in Afro-American Literature* (Urbana: University of Illinois Press, 1987).

11 Thomas C. Buchanan, *Black Life on the Mississippi: Slaves, Free Blacks, and the Western Steamboat World* (Chapel Hill: University of North Carolina Press, 2004). Daniel Usner is a prolific scholar in Colonial Louisiana studies but one of his most important

works remains *Indians, Settlers, and Slaves in a Frontier Exchange Economy: The Lower Mississippi Valley before 1783* (Chapel Hill: University of North Carolina Press, 1992).

12 When using the memoirs of former slaves, two articles are noteworthy for their caveats. Both, however, support the validity of these sources. See John W. Blassingame, "Using the Testimony of Ex-Slaves: Approaches and Problems" *Journal of Southern History*, Vol. 41, No. 4 (November 1975):473–492; David Blight, "The Slave Narratives: A Genre and a Source" *History Now: The Journal of the Gilder Lehrman Institute* accessed at https://www.gilderlehrman.org/history-by-era/literature-and-language-arts/essays/slave-narrative

13 W.T. Cowan, *The Slave in the Swamp: Disrupting the Plantation Narrative* (New York: Routledge, 2005):125.

14 Cynthia Gooch, "I've Known Rivers: Representations of the Mississippi River in African American Literature and Culture" Ph.D. Dissertation (University of Kentucky, 2019) accessed at https:uknowledge.uky.eddu/cgi/viewcontent.cgi?article=1125&context=English_etds; Henry Bibb, *Narrative of the Life and Adventures of Henry Bibb, an American Slave, Written by Himself* (New York, 1849; rpt. Echo Library, 2005):19.

15 Nell Irwin Painter, *Exodusters: Black Migration to Kansas* (New York: W.W. Norton and Company, 1992); Bryan M. Jack, *The St. Louis African American Community and the Exodusters* (Columbia: University of Missouri Press, 2008).

16 Mark T. Carleton, *Politics and Punishment: The History of the Louisiana State Penal System* (Baton Rouge: Louisiana State University Press, 1971); David M. Oshinsky, *"Worse Than Slavery": Parchman Farm and the Ordeal of Jim Crow Justice* (New York: Free Press, 1996); Tabitha LaFlouria, ""Under the Sting of the Lash": Gendered Violence, Terror, and Resistance in the South's Convict Camps" *Journal of American History*, Vol. 100, No. 3 (Summer 2015):366–384.

17 The 1927 Flood has been a popular subject for many, including John M. Barry, *Rising Tide: The Great Mississippi Flood of 1927 and How It Changed America* (New York: Simon & Schuster, 1997); Pete Daniel, *Deep'N As It Come: The 1927 Mississippi River Flood* (New York: Oxford University Press, 1977); Patrick O'Daniel, *When the Levee Breaks: Memphis and the Mississippi Valley Flood of 1927* (Charleston: The History Press, 2013); David Cohn, *Where I Was Born and Raised* (Notre Dame: University of Notre Dame Press, 1935). Environmental historian Christopher Morris contends, "The 1927 flood was, and by some measures remains, the nation's greatest natural disaster, remembered in photos, songs and film." See Morris, *The Big Muddy: An Environmental History of the Mississippi and Its Peoples from Hernando De Soto to Hurricane Katrina* (New York: Oxford University Press, 2012):165. For an excellent cultural history, see Susan Scott Parrish, *The Flood Year 1927* (Princeton: Princeton University Press, 2017).

18 Alan Lomax, *The Land Where the Blues Began* (New York: Pantheon Books, 1993); Paul Oliver, *Blues Fell This Morning: Meaning in the Blues* (1960; rpt. Cambridge: Cambridge University Press,1990); David Evans, *Big Road Blues: Tradition and Creativity in the Folk Blues* (Berkeley: University of California, 1982); *Nobody Knows Where the Blues Come from: Lyrics and History*, ed. Robert Springer (Jackson: University Press of Mississippi, 2006); Richard M. Mizelle, Jr., *Backwater Blues: The Mississippi Flood of 1927 in the African American Imagination* (Minneapolis: University of Minnesota Press, 2015.)

19 Richard Wright, "The Man Who Saw the Flood" *Eight Men* (1940; rpt. New York: World Publishing, 1961):102–155.

20 Joyce Marie Jackson, *Life in the Village: A Cultural Memory of the Fazendeville Community* (Washington: U.S. Department of the Interior, National Historical Park, 2003); Juliette Landphair, ""The Forgotten People of New Orleans": Community, Vulnerability, and the Lower Ninth Ward" *Journal of American History*, Vol. 94 (December 2007). Accessed at https://doi.org/10.2307/25095146. For an excellent overview of Hurricanes Betsy and Katrina, see the award-winning text by Andy Horowitz, *Katrina: A History, 1915–2015* (Cambridge: Harvard University Press, 2020).

21 The literature on Katrina and the 1927 Mississippi Flood is rich—a few notable works on Katrina include Douglas Brinkley, *The Great Deluge: Hurricane Katrina, New Orleans and the Mississippi Gulf Coast* (New York: Wm. Morrow and Company, 2006); Michael Dyson, *Come Hell or High Water: Hurricane Katrina and the Color of Disaster* (Davidson: Civitas Press, 2007); K. Wailoo and K.M. O'Neil, eds., et al., *Katrina's Imprint: Race and Vulnerability in America, Rutgers Studies on Race and Ethnicity* (New Brunswick: Rutgers University Press, 2010).

22 One of the best scholarly treatments of "Cancer Alley" remains Barbara L. Allen's, *Uneasy Alchemy: Citizens and Experts in Louisiana's Chemical Corridor Disputes* (Cambridge: MIT, 2003). To learn more about the individual grassroots organizations, consult individual websites. Robert D. Bullard, *Dumping in Dixie: Race, Class, and Environmental Quality* (Oxford: Westview Press, 1990); Rob Nixon, *Slow Violence and the Environmentalism of the Poor* (Cambridge: Harvard University Press, 2011).

23 Carolyn Merchant, "Shades of Darkness: Race and Environmental History" *Environmental History*, Vol. 8, No.3 (July 2003): 380, 384; Carolyn Finney, *Black Faces, White Spaces: Reimagining the Relationship of African Americans to the Great Outdoors* (Chapel Hill: University of North Carolina Press, 2014):54, 68; Mart Stewart, "If John Muir Had Been an Agrarian: American Environmental History West and South" *Environment and History*, Vol. 11, No. 2 (May 2005):139–162; Kimberly K. Smith, *African American Environmental Thought* (Lawrence: University Press of Kansas, 2007).

24 Scott Giltner, "Slave Hunting and Fishing in the Antebellum South" *To Love the Wind and the Rain: African Americans and Environmental History*, eds. Dianne Glave and Mark Stoll (Pittsburgh: University of Pittsburgh Press, 2005):21–37; *Black Nature: Four Centuries of African American Nature Poetry*, ed. Camille T. Dungy (Athens: University of Georgia Press, 2009):xxv. George Marion McClellan, "A September Night" (1916) accessed at https://poets.org/poem/september-night

25 Richard White, "Are You an Environmentalist or Do You Work for a Living?" *Uncommon Ground: Rethinking the Human Place in Nature*, ed. William Cronon (New York: W.W. Norton & Company, 1996):171–185; Gunther Peck, "The Nature of Fault Lines and Common Ground in Environmental and Labor History" *Environmental History*, Vol. 11, No. 2 (2006):212–238; Chad Montrie, *Making a Living: Work and Environment in the United States* (Chapel Hill: University of North Carolina Press, 2008); Stefania Barca, "Laboring the Earth: Transnational Reflections on the Environmental History of Work" *Environmental History*, Vol. 19, No. 1 (2014):3–27.

26 Joanna Dyl, "Transience, Labor and Nature: Itinerant Workers in the American West" *International Labor and Working Class History*, Vol. 85 (April 2014):97–117; Thomas G. Andrews, *Killing for Coal: America's Deadliest Labor War* (Cambridge: Harvard University Press, 2010).

27 Walter Johnson, *River of Dark Dreams: Slavery and Empire in the Cotton Kingdom* (Cambridge: The Belknap Press of Harvard University Press, 2013); Sven Beckert, *Empire of Cotton: A Global History* (New York: Vintage Books, 2014); Edward E. Baptist, *The Half Has Never Told: Slavery and the Making of American Capitalism* (New York: Basic Books, 2014); Paul Outka, *Race and Nature: From Transcendentalism to the Harlem Renaissance* (New York: Palgrave Macmillan, 2013).

28 Thomas D. Rogers, *The Deepest Wounds: A Labor and Environmental History of Sugar in Northeast Brazil* (Chapel Hill: University of North Carolina Press, 2010).

1

THE COLONIAL ERA

The River and Its Bayous

In 1784, the well-known maroon, San Malo, and his few remaining followers were captured by Spanish colonial law officers. After years of marronage and earlier failures by the Spanish, San Malo and his band were found in an area known as the "Land of Gaillarde," located between the Mississippi River and Lake Borgne, in which they had "occupied all the bayous that led to Lake Borgne." Contemporaries described the area where they were captured as a "swampy, forbidding region inhabited only by beasts, reptiles, and myriad insects." This last refuge of San Malo's was so treacherous that even some of his fellow maroons refused to follow him there. Once captured, San Malo and the other maroons—found "resting along the banks of the bayou"—were taken down the Mississippi River by way of pirogues as bystanders lined the river to watch, many offering gifts to the Spanish captors. The maroons were taken to New Orleans, where they stood trial on charges of inciting rebellion. All were found guilty and punished by death, except for one female member who claimed she was pregnant. For San Malo, capture meant the end of years subsisting in the Mississippi River swamplands, eking out a living in an environment so brutish that earlier colonial officials were deterred from finding him. Only with the theft of cattle and subsequent death of a planter were officials compelled to locate San Malo and his band.[1]

San Malo's plight became legendary in the African American community, as illustrated in the following ballad.

Alas! young men, come, make lament
For poor St. Malo in distress!
They chased, they hunted him with dogs,
They fired at him with a gun,

. . .

DOI: 10.4324/9781315617077-2

They hauled him from the cypress swamp.
His arms they tied behind his back,
They tied his hands in front of him;
and then they raised the gallows-tree.

They drew the horse—the cart moved off—
And left St. Malo hanging there.
The sun was up an hour high
When on the Levee he was hung;
They left his body swinging there,
For carrion crows to feed upon.[2]

Thus, for African Americans, San Malo's life was remembered as one of freedom, living in "cypress swamps," also known as cipiere, with an ending "on the Levee." (Actually, San Malo was not hung on the levee but in New Orleans. In a later slave conspiracy, however, the levee was the site where the heads of the "conspirators" were displayed.) His story reveals not only the place of freedom and resistance among enslaved people but also the role of the river and its environs in the African American community. In the Mississippi River Basin, the areas claimed by maroons were unwelcoming spaces to the planters and other members of the settlers' community. Maroons lived for days, months, and years in the lands bordering the plantations. These maroon communities could be found throughout the slave-holding South. One of the most well documented was in the Great Dismal Swamp (GDS), located in present-day states of Virginia and North Carolina. Recent archaeological findings revealed that generations of maroons lived in the GDS, with the children born in maroon families, never leaving these marginalized lands. In the Lower Mississippi Valley, for many maroons in the early years of French settlement, refuge was also found in nearby Native American communities. Others, like San Malo, traveled further inland into the bayous where the vegetation was thick and dense, perceived as inhospitable to the planters. For those who fled to this hinterland, they survived on turtles, fish, whatever game they could hunt, and roots and herbs. Other life-saving activities included harvesting Spanish moss from the plentiful cypress trees to sell to traders, working with area sawmills in the growing cypress lumber trade, and selling goods such as handmade baskets or garden vegetables in nearby markets. Given the maroons' skills as hunters and fishermen, wild fish and game were also probably sold.[3]

In testimony from members of San Malo's band—all prompted to run away due to mistreatment or wishing to join family members—they forged a living through a variety of means. Many such as San Pedro had families, as he lived with his wife and children in the outlying areas for three years, where "he raised provisions at Gaillard Land" for one year. Still another maroon, Philipe, who had joined different maroon groups for five years, survived on "wild fruit and fish." In other maroon testimonies, Goton, in describing Gaillard Land, said that they lived "on the root of an herb called China-smilax, pounded and made into flour which cooked." She also recounted a community of "many new and old" cabins, shared by all. Another testimony from San Malo's band said that they sold baskets made of willows. Their system for living in the hinterlands was elaborate, as they devised strategies to alert them if strangers tried to enter their camps. For example, earlier attempts by the Spaniards to capture San Malo were foiled when San Malo's band heard their approach through the rustling of interwoven cane palms and dry leaves that had been laid across the trail. San Malo had also selected campsites where "one had to wade through reeds in chest-high water," again a deterrent to Spanish officials. Although not a physical deterrent, San Malo had a sign posted outside his "Land of Gaillarde" site that translated "Woe to the white who would pass this boundary."[4]

In addition to sustenance and refuge, swamps were also theaters of resistance, an integral part of enslaved peoples's narratives of life in the Lower Mississippi Valley. In contrast to the colonists—whether French, Spanish, or American—the maroons' relationship with the swamps and borderlands was an intimate one as the cypress-studded swamps offered refuge, economic opportunities, and spaces to meet. Juxtaposed with these perceptions, however, were those of early nineteenth-century explorers, such as Timothy Flint. He described "sickly swamps" as "the haunts of fever, musquitos [sic], moccasin snakes, alligators, and all loathsome and ferocious animals, that congregate far from the abodes of man, and seem to make common cause with nature against him." In colonial Virginia, similar reactions to swamps had been voiced. In 1728, Colonel William Byrd II, when tasked to survey swamplands in the GDS, found the lands uninhabitable and described their challenges observing "the skirts of the Dismal towards the East were overgrown with reeds ten or 12 feet high, interlaced everywhere with strong bamboe-bryers, in which the men's feet were perpetually intangled." Byrd's description of extensive overgrowth resting on marshy bottomland reflected the perceptions of other early colonists. These views, whether swamps or wetlands in the north, were universally shared by the colonists and later, American settlers, and persisted up until the mid-1800s. But to enslaved people, the swamps and the unknown, unexplored plantation borderlands represented safety and freedom, offering sustenance and physical space free from oversight. The oft-quoted runaway enslaved person, Tom Wilson, best expressed this when he remarked, "I felt safer among the alligators than white men." Although the accounts are limited, another former slave from New Orleans disclosed how he ran away "to the woods," where he and a group of about thirty lived for almost one and one-half years. Again, swamps served several purposes for enslaved people; opportunities for freedom, sustenance, and sites of resistance as illustrated in San Malo's sign barring whites from the "Land of Gaillarde."[5]

San Malo's experience—though exceptional for the Lower Mississippi Basin for the length of time he survived as a maroon and the number of women and children in his communities—was not unique, as records document that enslaved people escaped to other surrounding swamps. In petitions brought before the

French governors, settlers contend that their cattle were being killed by slaves whom they claimed were "marooning." In one instance in 1737, an enslaved man named Geula was imprisoned for "frequent marooning." He had been living on a cypress tract, where he claimed that he had been mistreated while enslaved at one of the area's larger plantations. Colonial officials found him guilty with the inhumane, brutal sentence of cutting off both his ears and branding him. Other accounts include several maroons found on the German Coast, again accused of taking cattle and possessing firearms. Part of their survival included hunting deer and teal ducks. For the French, marronage was a constant threat provoking the passage of the 1724 Code Noir, Article 32 outlining the punishments for those attempting to escape bondage. The enactment of the law revealed the threat that marronage posed to the settlers. Through laws, such as this, newspaper accounts advertising rewards for runaways, and French administrative reports, there is ample evidence that marronage was a constant during the period of slavery on the North American continent beginning with the first Africans arriving to the Mississippi River Basin in the early 1700s and persisting through successive empires. As long as the institution of slavery remained, resistance through marronage along with other attempts to escape, all facilitated by the rich bayous and numerous waterways leading to freedom, would be part of the Mississippi River's history. In the Lower Mississippi Valley, marronage challenged the occupying powers as Spanish official records in 1781 noted that there was an ever-present "roving contingent of run-away slaves hiding in the swamps contiguous to New Orleans on both sides of the Mississippi River." The frequency of marronage and its associations with resistance, refuge, and economic activity also reflects a rival geography; a landscape where swamps as uninhabited space were appropriated by enslaved people representing another history of the colonial era. As settlers continued to arrive, the Lower Mississippi River Basin's landscape evolved with the architecture of riverfront plantations claiming fertile riverside tracts, backed by endless cypress swamps that were laced with numerous waterways navigating the dense vegetation. Enslaved people benefited from this plantation landscape—particularly in the early years of French settlement—which offered opportunities for escape with the thick swamp terrain lying beyond the planters' domain. But marronage and escape opportunities were only one facet of their history with the river and its bayous.[6]

But first an overview of the river's initial contact with Europeans beginning with Hernando De Soto, who in his efforts to find treasure was the first to record impressions of the Mississippi River in 1541. After De Soto's failed expedition, it would be another 125 years before settlement of the river valley began under France. The French—and their various commercial enterprises—remained until their defeat in the Seven Years' War and the signing of the Treaty of Paris in 1763 when Spain gained control of the territory. Spanish rule, however, was short-lived and after Napoleon's ascent in the early nineteenth century, France again governed the region until the fateful sale of the Louisiana Purchase to the United States in 1803. All sought control of the Mississippi River Valley, recognizing the area's agricultural potential and the river's commercial value as an inland highway for the shipment of goods that would reach a trans-Atlantic market. Given its location and physical characteristics, it is not surprising that the river was such a coveted acquisition for these empires. On the continent, the Mississippi River is unmatched in length, running for 3,705 kilometers with a river basin covering 4.76 square kilometers. The river's drainage area is the third largest in the world. Coinciding with the river's size was a topographical diversity ranging from the imposing, stacked bluffs lining the Upper Mississippi River that begins at Lake Itasca and ends near present-day Minneapolis, from which the river's landscape expands into the Yazoo-Mississippi Floodplain, ending at the confluence of the Ohio and Mississippi Rivers. Finally, the Lower Mississippi River, which by this time is home to its two major tributaries—the Ohio and Missouri Rivers—courses through a landscape featuring the rich southern Delta lands with their bordering swamps before emptying into the Gulf. The northern stretches of the river, or the Upper Mississippi River, runs through a less inviting environment for agriculture; however, in the early years of European settlement, it supported the fur trade. The Middle and Lower Mississippi Valleys, once cleared, would be home to some of the richest farmlands in the nation, generating millions of dollars for the planters who dominated the economy until the outbreak of the Civil War. For African Americans, the Lower and Middle Mississippi Valleys figured most prominently in their histories up until the Civil War. In addition to the Mississippi River, the major tributaries, including the Ohio, Missouri, and Red Rivers, played important roles in their histories. Where the European and American empires saw economic opportunities, for enslaved and free blacks, the riverine environments represented not only multiple images, including the conduit into slavery, the site of brutal labor regimes, but also resistance to the institution of slavery, and at times refuge and freedom.[7]

After France's and later Spain's efforts at colonization, in which neither nation retained the Louisiana Territory, from 1803 onward, the entire Mississippi River Basin has been solely within the boundaries of the United States, occupying a significant place in American history. Economically, the river's value is well documented as the Mississippi was the only major transportation artery for the interior with access to the Gulf. Up until the advent of the Civil War, southern states such as Mississippi and Louisiana, for example, enjoyed some of the nation's most lucrative economies. Trade on the river was robust, as the cotton empire of the Deep South supplied much of the world's cotton. For example, by 1860, there were 3,500 boats carrying cotton and other goods arriving at the New Orleans levee annually. Complementing the association with trade and commercial success, the river assumed cultural significance, becoming an American symbol, an exemplar of the frontier. To nineteenth-century Americans, the Mississippi waterscape rivaled the landscapes of Europe—contributing to a sense of nationalism. This imagery was memorialized in poems, nineteenth-century panoramas, artwork, and prose.[8]

However, the river's commercial value would never have been realized without the labor of enslaved Africans. Each empire—French, Spanish and

American—relied heavily upon enslaved and free black labor. By the mid-1800s, approximately 1.8 million slaves were growing more than two billion pounds of cotton that netted almost $220 million. Again, the Mississippi River was the vehicle for developing this area, accounting for an emergent global market in which the South produced most of the world's cotton production. The 1700s were the foundational years for the economic juggernaut that would become King Cotton and ensure Southerners' commitment to the institution of slavery, prompting the Civil War. Beginning in 1719 with New Orleans's selection as the site of French commerce in the lower Mississippi Basin, Africans were forcibly removed to the Mississippi River Valley where they rowed boats, built levees, worked as artisans, labored on plantations cultivating rice, indigo, tobacco, cotton, and sugarcane, and loaded and unloaded freight off incoming and outgoing ships, to name only a few examples of their labor. For African Americans, developing the city of New Orleans and subsequent efforts to enrich the French Empire is the first chapter of their history with the river. The work performed in the early years set the stage for a historical trajectory where labor remained constant, informing the history and collective memory of African Americans and the river.[9]

By the time the French claimed a foothold in the Lower Mississippi Valley with its gateway to the Atlantic world, their empire was already well established with territory as far north as Canada. Ambitious to expand, the French, with settlements as far north as present-day Illinois, recognized the river as the connector for their planned empire. From the onset, they sought to harness the river as officials stated that the ultimate goal was to forge a river for commerce and make the "entrance to the river navigable for all sorts of vessels and that will be one of the greatest advantages than can be procured for this country." As early as the 1720s, French officials in the Louisiana territory congratulated themselves with improvements to the river's mouth. Improving access to the robust trans-Atlantic trade was ongoing during the French presence, as reaching the site of New Orleans proved challenging to navigate. At times, boats coming from the mouth of the river had to be "hauled out of the water . . . and carried by the Indian or negro slaves until the river was reached again." In one correspondence to the Company of the Indies from early colonial officials, they emphasized the difficulty in navigating the river channel, observing the need "to maintain a good pilot." Undaunted, their visions of empire foresaw forts lining the river all the way from its upper reaches to the mouth. Rich trade networks would result in this new age of empire-building that dominated the sixteenth, seventeenth, and eighteenth centuries. But to achieve this, a permanent settlement was the goal following the voyageurs and their explorations in the lower territory in the early 1700s. By 1718, Jean-Baptiste Le Moyne, Sieur de Bienville, the colonial governor of the Louisiana colony, selected the site of the future city of New Orleans. Upon selection, one of the first requests for African labor can be found in Bienville's accounts where he is asking for "Negro slaves" to dig ditches and insure the future of New Orleans. Much of Bienville's tenure as governor consisted of protecting New Orleans from flooding and requesting slaves to perform the work. Later, the city instituted a corvee system of labor where planters were required to provide the labor of enslaved people for a set amount of time to build and maintain the city's levees.[10]

After the selection of the New Orleans site, the Company of the Indies, which had received a concession from the French government for the colony, began recruiting aspiring planters with promises of lucrative returns. The company remained in the Louisiana Territory until 1731, promoting the cultivation of crops such as tobacco, indigo, and later rice. In order to succeed—first in establishing New Orleans as a port city and second, producing and marketing crops—the planters and company officials required a dependable source of labor. Beginning with Bienville's request in 1719, pleas for "negroes" can be found on a regular basis in early French provincial records. In those first years, reports to the Company of the Indies included statements such as "All the colony is impatient to see some negroes, whom it great needs, arrive." Further, if these requests went unmet, colonial officials reported that early settlers would be returning to France. While African slaves served numerous colonial enterprises, officials recognized that with New Orleans' problematic low-lying location, the city was always subject to overflow, necessitating the maintenance and construction of levees from which enslaved labor was deemed critical. (The situation would worsen as upstream planters built their own levees, further corseting the river.) According to an early chronicler, Baron du Villiers, in 1719 Bienville wrote:

> it may be difficult to maintain a town at New Orleans; the site is drowned under half a foot of water. The sole remedy will be to build levees and dig the projected canal from the Mississippi to Lake Pontchartrain.

Dikes and levees were built, prompting one observer in 1720 to observe, "The Mississippi, overflowing more or less for six months of the year, renders New Orleans unpleasant as a place of sojourn. But at present, a great many slaves or negroes from Guinea are labouring to make it habitable." Compounding the annual flooding was another natural disaster, the yearly threat of hurricanes. Adding to these deterrents was the presence of several diseases such as pleurisy, pneumonia, as well as sub-tropical maladies. Still levees continued to be built, becoming an ever-present feature on the New Orleans landscape. In constructing the levees, once the Company of the Indies assumed control of the Lower Mississippi Valley, they began in earnest bringing Africans from the west coast of Africa to work in the Louisiana territory. In a document supporting Baron du Villiers' observations, one company official recorded expenditures for "A levee on the bank of the river in order to guarantee the city from overflow, built by the negroes of the company." Thus, for African Americans, their introduction to the Mississippi River would be mediated through their labor subduing the river.[11]

With the ongoing demands for African labor, the first African slave ships arrived in 1719 and by 1721, the Company brought "cargoes of Negroes" almost every year. Most were from Senegambia, although some came from the West

Indies. By 1729 when the majority of Africans had arrived during the colonial period, two-thirds were from Senegambia where the Company of the Indies had a concession. The African American population increased throughout the French colonial period—despite the number of diseases endemic to the area—in part due to their hunting and fishing skills and according to one scholar, their immunity to malaria. But mortality rates from the harrowing trans-Atlantic journey were often high. In one ship account during the early years of settlement, 280 Africans began the journey with only 180 surviving. From 1718 to 1731, out of a total of 7,000 Africans forcibly taken to Louisiana, only one-half survived. For those Africans who survived the ocean passage, they were then faced with the brutal process of being sold. For others, even if they survived the trans-Atlantic journey, they often arrived in New Orleans with debilitating illnesses. (In one 1729 report to the Directors of the Company of the Indies, blame for the poor health of the arriving Africans was based on too long a period at Goree under poor conditions in the slave prisons.) For example, in a 1727 report to the Company of the Indies, one ship had arrived with 266 blacks with many afflicted by "two diseases, some by a dysentery with a bloody flux and others by an inflammation on the eyes by which have been left one-eyed . . . and blind." In other instances, Africans suffered heavy losses from scurvy. According to one account, dated 1728, and addressed to the Directors of the Company of the Indies, colonial administrators relate a voyage from Senegal that began with 350 Senegalese. Three-hundred and forty-one survived the trip, only to have most of the passengers showing scurvy symptoms a few days after arrival. Still another 1720s' account detailed how one ship carrying 400 Africans from the prison at Goree arrived at the mouth of the Mississippi with only 260 alive. After their arrival, another 25 to 30 Africans died from scurvy. For those, however, ready to be sold, their next ordeal was the auction block, with one of the more famous auction houses in New Orleans, located a few blocks from the river. Doomed to a life enslaved in Louisiana Territory, sales to the planters involved inspections of their nude bodies, checks for disease and overall health—all this after a grueling Atlantic voyage. This process continued regardless of empires—French, Spanish, and the United States—up until the mid-nineteenth century. Thus, for African Americans, beginning in the 1720s, the Mississippi River and its environs was not only a site for labor but also the gateway to slavery and all that horrific institution entailed beginning with auction blocks poised near the riverfront.[12]

But for those who survived the trans-Atlantic passage, their labor-building levees produced a new riverine landscape as New Orleans could boast of a levee extending for a mile by 1727. The labor, however, of newly arrived African slaves was not limited to levee-building as one of the earlier ships from Africa included 40 men who left New Orleans and worked on boats traveling up the river to Illinois. As the French controlled all the land bordering the Mississippi River and while the majority of African Americans remained in the Lower Mississippi Valley, it can be assumed that others also made the trip upstream. Their skills as navigators must have been employed, as navigating the river upstream was treacherous. In one 1727 account from a Jesuit priest, he described his trip from New Orleans and encountering a "mass of floating trees which the river has uprooted, and which the current drags onward continually." The trees can become entangled, making these river stretches difficult and combined with a swift current, "if the pirogue be driven against the floating trees it would immediately disappear and be swallowed up in the water." Unharnessed, the river was a force to be reckoned with as explorers and later river pilots recounted navigating the ever-changing waterway. As the Company brought more Africans to the territory, their history with the river included laboring as river pilots and oarsmen.[13]

Thus, by the early part of the eighteenth century, with the recruitment of more French entrepreneurs and settlers to this promising landscape, black labor expanded to support a host of jobs, including the construction of levees and ditches, agriculture, river pilots, and eventually, the cypress lumber industry. Still, in those first years, when the majority of African Americans lived within a boat ride of New Orleans, most of the work consisted of subduing the unruly Mississippi Basin with levees, ditches, canals, and docks to facilitate commerce. This would soon change, however, as French administrators sought a staple crop to produce for export. But to be successful, the colonists were insistent that they needed black laborers. So, by 1734, "colonists from New Orleans and Mobile [were] making urgent appeals for slaves offering for them as much as 1,400 livres apiece." As long as French colonial administrators were overseeing the territory, they continued to see this need and according to Vaudreuil, "the advancement of the province absolutely depended upon an augmentation of slave labor." In one report, it was indicated that the colonists would return to France if more "negroes" were not sent. A constant refrain, in 1752 the demand for labor to support agriculture, was still being voiced in colonial administrative accounts. In all the requests, the assumption was that any major agricultural undertaking would be impossible without African labor. And so, throughout the 1700s, the landscape of the Mississippi River Valley underwent major changes as African American labor was expended to grow crops and harness a river. The narrative of the Mississippi River and Africans begins with their forced labor to contain the river, enabling empires to prosper within this riverine environment.[14]

As the need for levees to protect New Orleans from periodic flooding persisted, particularly in the early years to "guarantee it for the future," digging ditches and building levees were a part of eighteenth-century life for African Americans. In planning future levee improvements, French officials relied upon the promise of more "negroes." As mentioned earlier, a corvee system evolved and as settlement grew, riparian landowners were expected to provide their enslaved' labor for public works, which included the building of levees. Statute required each planter to contribute 30 days' labor to public works in New Orleans. But levees were also required to front the homes of the planters who lived along the Mississippi River, which again were usually built by the labor of enslaved people. (This was in addition to the statute that required 30 days' labor for public works.) These riverfront levees would eventually line the Lower Mississippi Valley as the planters expanded upriver. Once the Company of the Indies assumed control of the Louisiana Territory, they opened settlement up from the mouth of the river to the rich farmlands bordering the river up to Natchez. These lands were fertile, promising healthy profits when tobacco and indigo were planted. The Company of the Indies, in cultivating the rich lands around Natchez, encouraged growing tobacco and by 1725, remarked that "the harvest is abundant." But the promise of a stronger market was always accompanied by the same caveat "The plantation of the Company would succeed well if there were good Negroes." Still tobacco was considered a promising crop, as one observer remarked in 1727,

> The French settlement at Natchez is becoming important. Much tobacco is grown there which is considered the best in the Country. The situation of the town is very high; from it the Mississipi [sic] can be seen winding as if in an abyss; there are continuous hills and valleys. The land of the concessions is more level and of better quality.

In another French colonial account, the author claimed that the tobacco grown at Natchez was "even preferable to that of Virginia or St. Domingo." The planting of staple crops, such as tobacco, cotton, and sugarcane, ensured the presence of African Americans in the Lower Mississippi Valley as settlers and colonial officials believed that enslaved labor was imperative to succeed.[15]

So as African American labor—through agriculture and levee-building—contributed to empire-building in the lower Mississippi Valley, African Americans became critical to the French colonial economy. Black labor filled economic niches, as the importance of enslaved labor in other venues was observed. According to company records in 1728, they were "placing negroes as apprentices with all the workmen who we think are good and honest men." In this same correspondence, mention was made that this practice should have begun earlier while agriculture was still one of the areas where labor was needed. By 1734, however, black slaves worked in various trades as Bienville reported that "Sieru Dubreuil, the contractor of all the King's works, has trained negroes in all kinds of trades he employs only very few French workmen . . ." For example, in one 1727 Superior Council record, a locksmith in New Orleans committed to teaching his trade "for three years to a slave apprentice." The apprentice would be selected from the next "slave ship." Another official complaint referenced the work of a runaway who had been hired to work as a bricklayer, tiler, and roofer. In one of Bienville's earlier requests for slave labor, he not only wanted Africans to help in building levees but also asked for Africans who were familiar with growing rice. In Senegambia, where the Company of the Indies held a concession, rice cultivation was common, although millet was the more common crop. Both foods, however, prepared the slaves "from the Senegambia hinterland . . . for the kind of labor required in the Mississippi Valley." Still in the early years, rice became important for a local market. As early as the early 1720s, one provincial report discussed a concession near New Orleans that "yielded 600 quarters of rice against 14 quarters sowed." The author also noted that 46 "negroes" worked on the concession. In another glowing report from the colonial administration, the rice crop was "abundant," as the official claimed: "I cannot believe that there is a country in the world that is more suitable for it." The same report, however, conceded that rice would not be a rich cash crop for France but served the subsistence needs of settlers and "negroes."[16]

Returning to their work on the river, African American slaves also labored on the river piloting boats downstream where the river proved as challenging as upstream travel. As early as 1727, one colonial account records the work of 55 "negroes" meeting ships at the mouth of the Mississippi. Entering the river could be challenging as pilots negotiated sandbars and shallow depths for part of the year, requiring navigational skills that many of the slaves undoubtedly already possessed from working on the Senegal and Gambia Rivers. In another French provincial correspondence, the author discusses how a recent shipment of "negroes" have been assigned work. In his words, "fifty to work on the levee" and "There are three of them who the captain says are good sailors who are assigned to that." Still another reference to having enslaved people employed as sailors is found in a report from one of the provincial governors to the Directors of the Company of the Indies (also known as the Law Company) in 1724. He observed that two negroes from Suratte were "now good sailors" and suggested that "those who you think suitable be retained in order to make sailors of them." The Company framed this labor in economic terms, stating "we shall diminish the naval expenses every day by making only blacks and a few white men, sailors." Other references to the river areas, during the first years of the French presence, observe newly arrived Africans working along the river, piloting flatboats, and navigating an unruly river. Again, for many enslaved people this was a familiar environment as more than half of the first Africans, from 1719 to 1731, brought to the southern Mississippi Delta were from Senegambia where the Company of the Indies had a concession. The Africans in the region lived along the Senegal River Valley, familiar and knowledgeable with a riverine environment. As a result, their skills as boatmen were valued and their presence was more noted throughout the eighteenth century when the population of enslaved African Americans grew substantially. By 1797, African Americans represented an estimated 2% of the labor force working on the river. Though initially small, their numbers grew as steam dominated the river in the nineteenth century. More importantly for the African American community, working in river trades allowed more freedom and opportunity to communicate with others than those restricted to lives on plantations. River trades would continue to be associated with greater freedom in the nineteenth century.[17]

Further evidence of the role that African labor played in the early years can be found in census records documenting the population of the lower Mississippi River Valley. In 1712, only ten blacks lived in the Lower Mississippi Valley. Those numbers grew, however, with the arrival of more ships in 1719 when the first slave ships with Africans knowledgeable about growing rice arrived. By 1746, Louisiana

Territory had grown with a white population of 3,200 and 4,730 blacks. Most of the African Americans were from West Africa with a small group from the West Indies. By 1754, when the map of Louisiana was being redrawn with the Seven Years' War, two-thirds of the slave trade originated in Senegambia. With the large number of enslaved people coming from West Africa, Gwendolyn Midlo Hall, the well-known historian of colonial Louisiana, observed that "Louisiana [was] thoroughly Africanized in the early years." The African American culture that emerged in the Lower Mississippi Valley differed from that of other areas of the slaveholding South. With the presence of the expansive trans-Atlantic trade network, New Orleans was one of the most cosmopolitan cities on the North American continent. This urbane, cosmopolitan culture also shaped a cosmopolitan African American society in New Orleans derived in part from a slave society that was "international, interconnected and itinerant." New Orleans' connections with the Atlantic world of commerce that resulted in a metropolitan, fluid society enriched the collective memory of African Americans and their linkage with the Mississippi River, a linkage still present today.[18]

However, despite finding a profitable crop in tobacco, the Company of the Indies struggled during its tenure, never realizing the colonial riches they had envisioned. The Natchez Revolt of 1729, led by the Natchez tribe with many African slaves as allies, signaled the end of the company. At the time of the revolt, the Natchez settlement was occupied by 200 French men, 82 French women, 150 French children, and 280 black slaves. After the revolt, more than 200 French settlers were killed, leaving the settlement in ruins. When French officials from New Orleans sent soldiers to Natchez to retaliate, former black slaves allied with the Natchez to ensure they would not be defeated. In this instance, slaves and Indians were allies but this was not always the case. In the first years of the French colonial government, slaves were known to run away to neighboring tribes. By some accounts, Native Americans taught the slaves how to survive in nearby swamps whereas Africans taught the Native Americans how to fight against the French. Colonial administrators, however, realizing how their alliance could threaten French settlement, worked hard to keep the two groups divided, succeeding in many instances. As early as 1728, this concern was voiced when Étienne de Perier, the fifth governor of the Louisiana colony, wrote about his apprehensions regarding nearby Native Americans who had been enslaved. He observed, "these Indian slaves being mixed with our negroes may induce them to desert with them, as has already happened, as they may maintain relations with them which might be disastrous to the colony when there are more blacks."[19]

After the Natchez Revolt, the French government retained its hold over the colony until 1763 when France ceded its territory to Spain and Britain with the conclusion of the Seven Years' War. During the intervening years, however, the French government neglected its Louisiana holdings, an "abandonment period" for the colony. As a result, the economy diversified and a small-scale economy emerged that engaged settlers, slaves, and Native Americans. The economy that scholars labeled a "frontier exchange economy" was defined as "networks of cross-cultural interactions through which native and colonial groups circulated goods and services." For enslaved people, the economy allowed more freedom, which changed when Spain gained control in 1763. Upon the arrival of the victorious Spanish, a more circumscribed plantation economy emerged and the territory evolved from a "society with slaves to a slave society." Contributing to the change were events on the nearby Caribbean islands, including the Santo Domingo Revolt, resulting in the arrival of more planters and enslaved people to the Lower Mississippi Valley. Under the tutelage of the Spanish, the displaced Caribbean planters developed a thriving plantation economy, significantly affecting the lives of African American slaves. The importance of the Mississippi River became increasingly apparent as commerce increased and by the end of the Spanish era "commerce doubled every few years." By the 1790s, the lucrative crops that the Company of the Indies sought became a reality under Spanish colonial rule with a "full-fledged cotton and sugar boom underway." Again, the labor of African American slaves was critical for the successful cultivation of their crops. These were foundational years for the "workscape" that would be associated with the Mississippi Valley. By the 1860s, cotton plantations lined the Mississippi River with sugarcane plantations in the lower valley. Images of the two—plantation workscapes and landscapes—became part of Southern mythology, enshrined in the collective memories of blacks and whites.[20]

The Spanish, however, developed other economies, and one of the most important was the harvesting of cypress used for the containers that shipped sugar. Cypress trees, valued because they do not deteriorate or rot, provided an ongoing industry for whites and blacks. But even before the development of the lumber industry, the location of the riverfront plantations spurred an African economy connected with the cypress swamps. In stimulating an agricultural economy during the first decades of the eighteenth century, the first planters were given tracts of land that bordered the river, beginning around New Orleans and eventually reaching Natchez. Behind the cultivated tracts, known as arpents, lay the cypress swamps. In the area that stretched from Lake Pontchartrain to the mouth of the river, known as Bas du Fleuve (the area known as Gaillardeland), were the cypress swamps that supported San Malo and countless other maroons. French planters grew profitable crops on these riverfront tracts, whereas slaves who ran away to the swamps found work in the sawmills, supplementing a living that relied upon what they harvested from the surrounding swamps. Although challenging to enter and navigate swamp terrain, in the words of one colonial official, "It is a most propitious land for the maintenance of human life because of sweet potatoes, because of the great abundance of forest products, of much fish and shell fish, and abundant wildlife." In this instance, white and black narratives converged as the cypress swamps were economic mainstays for both. The swamps offered resources to market as well as to sustain populations.[21]

As the case of San Malo illustrated, the swamps also supported maroon communities with increasing numbers during the Spanish colonial empire's tenure in the Lower Mississippi Valley. Although San Malo and his band were the most well-known maroon communities during the Spanish era, other maroon communities were becoming established in reaction to the circumscribed plantation economy imposed by Spanish colonial officials. Marronage was distinguished between two types of escapes—long and short term. The first, grand marronage referred to the slave who ran away for long periods of time or even permanently. The second, petit marronage described shorter periods of escape or "periodic slave truancy." In addition to these distinctions, Slyvia Diouf, another scholar studying marronage, added two more defining characteristics of marronage. Diouf distinguished between "borderland maroons" and "hinterland maroons." The first describes those maroons who chose to remain nearby while the latter settled farther away.[22]

Whether long- or short-term, borderland or hinterland, by 1797 the incoming Spanish governor of Louisiana Territory complained that "cimarrones [maroons] were everywhere and something needed to be done." In response to the perceived threat, the governor created an agency, with the sole task of locating runaways. In the words of one historian, while there had always been maroons in the Lower Mississippi Valley, by the 1780s, the planters became alarmed as the maroons had "never organized themselves to the extent that maroons in the 1770s demonstrated." The fear was that an "alternate society" was being established. Although their numbers might have increased, the survival tactics for the maroons remained the same throughout the century. In order to sustain themselves, maroons relied upon edibles found within the swamps, provisions taken or supplied by nearby plantations, income generated from the cypress swamps, and the production of handmade goods, such as baskets. Even the taking of cattle became a common occurrence for maroons struggling to remain free. An intricate network evolved, as those slaves remaining on the plantations interacted with the maroon community, sometimes providing food or selling their goods at nearby markets. But underlying these networks was the maroon's sophisticated knowledge and skill in harvesting their resources. The maroons knew their landscape, a maroon landscape "whose several parts were connected by secret paths, discreet trails and waterways navigated under cover of night and whose outer, intangible limits reached dangerously into the plantations and cities." These areas—free from planter oversight—were mastered by the maroons. Their ecological knowledge coupled with the underground economy that evolved and included the production and sale of goods, working with nearby lumber mills, and relying upon relationships with enslaved people resulted in communities that sometimes lasted years. But maroon communities are even more impressive when considering the punishments meted out to maroons if captured. Brutal punishments ranging from flogging to branding to death, to name a few, remain part of the history of marronage. For those who lived along the river, violence will always be associated with the riverine environment. But the drive to be free—whether prompted by a planter's abuse or the need to join family—also became part of the river's history as African Americans appropriated and mastered spaces deemed uninhabitable to the planters.[23]

By the 1790s, another defining moment marked the long history between African Americans and the Mississippi River when news of the French Revolution reached enslaved and free blacks in the Mississippi Valley.[24] Infected by the revolutionary rhetoric emanating from the Revolution along with the abolition of slavery in the French Caribbean and the successful Haitian Revolution, enslaved people in the Pointe Coupee region, located by present-day Baton Rouge, planned a revolt. Later known as the Pointe Coupee Conspiracy, free and enslaved blacks, along with a few whites, met covertly in the cipriere, the cypress swamps, behind one of the main plantations, and strategized their attack. The plan was to set a building on fire, causing nearby planters to rush over to extinguish it. Once the planters arrived, the conspirators planned to ambush them, followed by attacks on nearby plantations. Before enacting the plan, the group was betrayed and brought to trial. Fifty-seven enslaved people and three whites were found guilty. Twenty-three of the enslaved people were sentenced to death by hanging. After the hangings, in an effort to deter any future revolts, colonial officials had them decapitated with their heads placed upon posts lining the river. With this grim show reflecting the cost of capture, the river again served as a reminder of the violence associated with slavery and the Lower Mississippi Valley. Another imaginary, however, was the role of the cypress swamp or cipriere, in facilitating resistance. The cipriere proved again to be an appropriated space for the African American community where resistance and identity could thrive. Both imaginaries persisted into the next century, as the river served as the conduit into the bowels of slavery and all the violence associated with the institution while alternately a means to escape whether through its swamps or the river, itself.[25]

By the end of the colonial era, two Mississippi River geographies emerged. On the surface, there was an inviting landscape to colonial empires, intent upon producing profitable agricultural goods with a waterway to the Gulf Coast, insuring trans-Atlantic trade. But within that landscape were challenges to be met such as navigating an inhospitable river or clearing lands and dense, forbidding swamps. Competing with this geography was another landscape, intricately tied to the river and its bounty. In this one, a world comprising waterways coursing through cypress swamps, inhabited by fish and alligators, supported by roots and herbs was becoming known and mastered by enslaved people. Both geographies existed up to the Civil War, although the second one would shrink as more land was cleared for settlement. Still, the landscape of cypress swamps remained in the historical memory as the "Orphan's Song," revealing lives dependent upon the swamp.

Little ones, without father,
Little ones, without mother,
What do you do to earn money?
The river we cross for wild berries to search;
We follow the bayou a'fishing for perch
And that's how we earn money.

....
Palmetto we dig from the swamp's bristling stores
And sell its stout roots for scrubbing the floors;

....
For making tea we collect sassafras,
For making ink, we collect pokeberries,

....
We go to the woods cancos berries to fetch;
And in our trap cages the birds we catch.[26]

Perhaps not as well known as the first historical landscape, the "Orphan's Song" illustrates a different type of kinship with the river's environs—a knowledge to be prized by African Americans. For African Americans, the knowledge represented a mastery of their surroundings and within that mastery another form of resistance to slavery. Knowing the river and its bayous could result in at best, freedom and at the least, resistance to the seeing world of the planter. Still another landscape emerged during the eighteenth century, the workscape of black labor. From the onset of empire-building in Mississippi River Valley, the history of the river was entwined with black labor. To the French, making New Orleans livable required constant levee-building and ditch digging with African Americans providing the labor. But once New Orleans was sustainable, serving as the gateway to Atlantic world commerce, the empire demanded profits from rice, indigo and tobacco, and later sugarcane and cotton. Again, African Americans supplied the labor in these rich, productive Delta farmlands. The river was always the backdrop in this multidimensional history and persisted in this ambiguous role of refuge as well as oppressor.

By the time of the American Revolution in 1776, the institution of slavery was entrenched along the Gulf Coast and Lower Mississippi Valley. But the lives of free blacks reflected a diversity in the pre–Revolutionary period as slavery evolved throughout the 1700s. Black laborers could be found in a number of trades with cartage trades allowing a degree of freedom for boatmen and canoers. The Revolution also offered a brief window for opportunities to escape slavery, as Great Britain sought the support of slaves in her battle against the colonists. But after the Revolution, the Lower Mississippi Valley began its shift to a more permanent plantation economy based on cotton and sugar that would also affect black lives. In retrospect, the colonial era would prove to be less constricting for black laborers as the nineteenth century crystallized into the "slave society" that began with the Spanish. In the words of one scholar, the Lower Mississippi Valley changed "from a society with slaves to a slave society." So by the end of the eighteenth century, the Mississippi Valley was a different region than the one Bienville encountered in the early 1700s, supporting new populations and commercial ventures. In 1712, ten blacks lived within the Lower Mississippi Valley. But those numbers grew with the arrival of slave ships starting in 1719. At one time, as mentioned earlier blacks outnumbered whites when in 1746, the Louisiana Territory had grown with a white population of 3,200 whites and 4,730 blacks. New Orleans, though not free from devastating floods, was, nevertheless, a city, rich in commerce, serving as a gateway to the Atlantic World and the interior of the continent. Levees and dikes surrounded the city, and one scholar described it as a city "retrieved from a swamp." By the early 1800s, the city and the Mississippi River Valley showed signs of the economic power they would soon possess. Integral to all of this were the lives of African Americans. Through black labor, levees were constructed for protecting New Orleans, commerce was robust, cypress swamps were harvested, and crops were grown. Along with that forced labor, a different narrative of the river emerged. Instead of presenting challenges, the swamps were considered forbidding spaces but havens and spaces to harvest natural resources, allowing for a degree of autonomy. The river was not only a space for commerce but also a means of escape and an outlet for commerce. So, for enslaved people, the possibility of escape by means of the river and swamps produced a riverine landscape that incorporated flight.[27]

In conclusion, the early history of enslaved and free blacks living in the Mississippi River Valley reveals a relationship with the river that rests upon several themes. After being forcibly removed from homes and communities in Senegambia and other west African civilizations, the African Americans' initial experiences with the river were mediated through their labor. During the colonial era, the work associated with the river—ranging from levee work to agricultural to lumbering in cypress swamps—produced a laboring landscape for enslaved and free blacks. But the associations were multidimensional and extended beyond labor as the river and swamps also represented refuge, sustenance, resistance, and freedom. Even if enslaved, laboring on the river often allowed more freedom than agricultural work. The laboring landscape that included building levees or working in sugarcane and cotton fields became part of a collective memory for African Americans. The nineteenth century with an entrenched, wildly successful plantation economy built upon these memories.

So, by the time of the Louisiana Purchase in 1803 with the United States' claim to the entire Mississippi River Basin, the contours of African Americans' experiences with the river had been established. Unlike the perspectives of French, Spanish, or American planters, for African Americans the riverine landscape was a laboring landscape. For African Americans, their history with the river was framed through the lens of labor, refuge, resistance, and sustenance. Labor, whether directly on the river, building levees, working on a riverfront plantation, or lumbering in a cypress swamp, was the point of entry for many African Americans. But these formative years included other intersections with the river informing a collective memory, rich in cultural associations by the twentieth century. More abstract than the laboring landscape, the river represented a loss of freedom, whether the end-stop of the trans-Atlantic passage or the auction blocks bordering New Orleans and Natchez riverfronts. Conversely, the riverine environment represented freedom through swamplands that offered sustenance and the means for a small-scale economy. The persistent efforts of the French and Spanish officials to deter marronage testify to the swamps as refuge. Paralleling the increased vigilance was the hardening of a plantation economy, derived from the rich alluvial soils of the Mississippi River Basin, not only resulting in more circumscribed lives for African Americans but also solidifying the differing riverine geographies. The bayous behind the plantation homes became spaces free from oversight—a distinction with more ramifications during the antebellum period. The uncultivated, untamed lands became those spaces for covert meetings, worship, and escape.

But other narratives were being formed during these years as the ongoing efforts to harness the river through levees, for example, mirrored the subduing of a black slave labor force. These were the foundational years for a long association with levees. The back-breaking work of levee-building continued into the early twentieth century as the Mississippi River refused to stay within its banks and again, African Americans constituted a disproportionate part of the workforce charged with protecting Delta lands and Mississippi River Valley cities. For African Americans, their connection with ditch-digging and levee-building would endure for more than three centuries into the twentieth century, a staple in their history of the river. By the outbreak of the Civil War, more than 2,000 miles of levee work lined the Mississippi River. A collective memory incorporating levees evolved as levees were linked to numerous imageries including the site of lynchings, barbaric levee camps, the Great Flood of 1927, and also the site of New Orleans Sunday markets and socialization. By the twentieth century, blues musicians captured these memories as they lamented the back-breaking work and harsh conditions in the levee work camps and the ever-present threatening floodwaters. During the next century, the antebellum era witnessed a further crystallization of these associations with the river as African Americans lived and worked along the river and cypress swamps. This riverine environment contributed to a collective memory with the river and its multiple perspectives ever informing and shaping identity.

Notes

1 Gilbert C. Din, ““Cimarrones” and the San Malo Band in Spanish Louisiana” *Louisiana History: The Journal of the Louisiana Historical Society*, Vol. 21, No. 3 (Summer 1980):247–248. When the Spanish controlled the Louisiana territory, they referred to the maroons as cimarrones. Numerous works examine San Malo and his band, noted in the Introduction. These texts draw upon Spanish Judicial Records, providing a comprehensive overview of the Spanish and French colonists’ concerns and San Malo’s capture.

2 George Cable, “Creole Slave Songs” *Century Magazine*, Vol. XXXI, No. 6 (April 1886) accessed at https://babel.hathitrust.org/cgi/pt?id=nc01.ark:/13960/t8tb3t099&view=1up&seq=7

3 One of the most comprehensive histories of life in the GDS can be found in Daniel Sayers, *A Desolate Place for a Defiant People: The Archaeology of Maroons, Indigenous and Enslaved Laborers in the Great Dismal Swamp* (Gainesville: University Press of Florida,

2014); see Gwenolyn Midlo Hall, *Africans in Colonial Louisiana: The Development of Afro-Creole Culture in the Eighteenth Century* (Baton Rouge: Louisiana State University Press, 1992) for a discussion about maroons living in Native American communities; Shannon Lee Dawdy, *Building the Devil's Empire: French Colonial New Orleans* (Chicago: University of Chicago Press, 2008):86.

4 Index to Spanish Judicial Records of Louisiana XL May–July 1981, 26 May 1781, *Louisiana Historical Quarterly* (1933):517, 519–520; 17 March 1783: 846–849, 864; Din, "Cimarrones," 254; Hall, *Africans in Colonial Louisiana*, pp. 18207, 213.

5 Timothy Flint, *The History and Geography of the Mississippi Valley*, Vol. 1, 3rd ed. (Boston: Carter, Hendee and Col, 1833):41; William Byrd, *Description of the Dismal Swamp and a Proposal to Drain the Swamp (1728)*, https://www.loc.gov/resource/lhbcb.22884/?sp=16.; Tom Wilson, *Albion* (Liverpool) 20 February 1858 in *Slave Testimony: Two Centuries of Letters, Speeches, Interviews and Autobiographies*, ed. John W. Blassingame (Baton Rouge: Louisiana State University Press, 1977):340. For a historical perspective of how wetlands have been perceived in the United States, see Ann Vileisis, *Discovering the Unknown Landscape: A History of America's Wetlands* (Washington: Island Press, 1997).

6 "Records of the Superior Council of Louisiana" *Louisiana Historical Quarterly*, Vol. 22 (April 1939):548–549; Vol. 5 (April 1922): 246–247; Vol. 5 (July 1922): 386–387; Vol. 19 (July 1936): 768–770; *Spanish Judicial Records*, 26 May 1781:516.

7 For a general history of the Mississippi River, see Paul Schneider, *Old Man River: The Mississippi River in North American History* (New York: Henry Holt and Company, 2013). Other works include John O. Anfinson, *The River We Have Wrought: A History of the Upper Mississippi River* (Minneapolis: University of Minnesota Press, 2003); Christopher Morris, *The Big Muddy: An Environmental History of the Mississippi and Its Peoples, from Hernando De Soto to Hurricane Katrina* (Cambridge: Oxford University Press, 2012); Mikko Saikku, *This Delta, This Land: An Environmental History of the Yazoo-Mississippi Floodplain* (Athens: University of Georgia Press, 2005). These are only a few of the numerous scholarly works on the river.

8 For an early history of commerce on the Mississippi River, see E.W. Gould, *Fifty Years on the Mississippi or Gould's History of River Navigation* (St. Louis: Nixon-Jones Printing Co., 1889). A more detailed discussion of the Mississippi River's cultural symbolism can be found in Dorothy Zeisler-Vralsted, *Rivers, Memory and Nation-Building: A History of the Volga and Mississippi Rivers* (New York: Berghahn Books, 2014), particularly Chapter 2.

9 In recent years, scholars have revisited the institution of slavery in the context of global capitalism. Their findings indicate that the cotton industry in the Mississippi River Valley was a very lucrative market that contributed to the South's emergence as one of the most profitable economies in the nineteenth century. See W. Johnson, *River of Dark Dreams: Slavery and Empire in the Cotton Kingdom* (Cambridge: Harvard University Press, 2013); S. Beckert, *Empire of Cotton: A Global History* (New York: Vintage Books, 2014); E.E. Baptist, *The Half Has Never Been Told. Slavery and the Making of American Capitalism* (New York: Basic Books, 2014).

10 *Mississippi Provincial Archives, 1701–1729* (hereinafter *MPA*) 3 vols. (Jackson, 1929), II, 658; Gould, *Fifty Years on the Mississippi*, 180; *MPA*, III, 388; Elizabeth Fussell, "Constructing New Orleans, Constructing Race: A Population History of New Orleans" *Journal of American History*, Vol. 94 (December 2007):846–855 may be accessed at http://archive.oah.org/special-issues/katrina/Fussellefd6.html?link_id=sco_early-immig Before the settlers were required to furnish labor by their enslaved black laborers, one of the French administrators complained about their lack of attention to the much-needed levees. In his official correspondence, he lamented "These levees were begun by squads of negroes, but this labor was neglected as the inhabitants owned very few negroes and did not spare them for this work." He then asks for a decree requiring the contributed labor. See *MPA*, I, 116. For an official reference to the decree, see *MPA*, II, 589.

11 *MPA*, II, 402, 492. For introductions to the founding of New Orleans, see Shannon L. Dawdy, *Building the Devil's Empire: French Colonial New Orleans* (Chicago: University of Chicago Press, 2008) Ned Sublette, *The World that Made New Orleans: From Spanish Silver to Congo Square* (Chicago: Lawrence Hill Books, 2008); Baron Marc du Villiers, "A History of the Foundation of New Orleans (1717–1722)" *Louisiana Historical Quarterly*, Vol. 3, No. 2 (April 1920):194–195; Albert E. Cowdrey, *Land's End: A History of the New Orleans District, U.S. Army Corps of Engineers, and Its Lifelong Battle with the Lower Mississippi and Other Rivers Wending Their Way to the Sea* (Washington: United States Army Corps of Engineers, 1977):1; Usner, "From African Captivity to American Slavery" *Louisiana History*, Vol. 20, No. 1 (Winter, 1979):183; Ira Berlin, *Many Thousands Gone: The First Two Centuries of Slavery in North America* (Cambridge: Harvard University Press, 1998):82–83.

12 N.M. Surrey, "The Commerce of Louisiana During the French Regime, 1699–1768," Ph.D. Dissertation, Columbia University (1916):232, 237, 243; Hall and Charles Henry Rowell, "American Tragedy: New Orleans Under Water" *Callaloo*, Vol. 29, No. 4 (2006):1049; Hall, Africans in Colonial Louisiana, p. 73, 203; Usner, "From African Captivity to American Slavery," p. 28, 34; *MPA*, II, 638, 547, 575, 620. Although the African American population grew during the French colonial period, respiratory and intestinal illnesses, to name a few, affected their numbers. See Daniel H. Usner, Jr., *Indians, Settlers, & Slaves in a Frontier Exchange Economy: The Lower Mississippi Valley before 1783* (Chapel Hill: University of North Carolina Press, 1992):34.

13 Cowdrey, *Land's End*, pp. 1–2; "Letter from Father du Poisson," *Travels and Explorations of the Jesuit Missionaries in New France, 1610–1791*, Vol. LXVII, ed. Reuben Gold Thwaites (Cleveland: The Burrows Brothers Company, 1900):289.

14 Berlin, *Many Thousands Gone*, p. 84; Surrey, "The Commerce of Louisiana," pp. 237, 243; *MPA*, II, 402, 658.

15 *MPA*, II, 408; *MPA*, II, 591–592; Cowdrey, *Land's End*, p. 1–2; *MPA*, II, 492, 565; *Travels and Explorations of the Jesuit Missionaries in New France*, Thwaites, ed., p. 311; M. Dumont, *Some Abstracts from the Historical Memoirs of Louisiana, Including the Most Interesting Events from 1687 to the Present* (1753) accessed at https://www.gutenberg.org/files/9153/9153-h/9153-h.htm#book-II-chapter-XII

16 *MPA*, II, 599; *MPA*, III, 779; James T. McGowan, "Creation of a Slave Society: Louisiana Plantations in the Eighteenth Century" Ph.D. Dissertation, University of Rochester (1976):138, 180; *Louisiana Historical Quarterly*, Vol. 1, p. 100; 5 October 1727, in Superior Council, *LHQ*, IV (1921):230; 5 (April 1922):244. *MPA*, II, 519. For the groundbreaking work regarding the transfer of African knowledge about rice cultivation to slave-holding societies, consult Judith A. Carney, *Black Rice: The African Origins of Rice Cultivation in the Americas* (Cambridge: Harvard University Press, 2001). Carney, in her study of slavery in Georgia and South Carolina, contended that one of the benefits for African slaves introducing rice crops was the opportunity to negotiate the conditions of their enslavement. Since this publication, critics of the "black rice" hypothesis have emerged with compelling arguments, such as the one set forth by D. Eltis, P. Morgan, and D. Richardson, "Agency and Diaspora in Atlantic History: Reassessing the African Contribution to Rice Cultivation in the Americas" *American Historical Review*, Vol. 112, No. 5 (December, 2007):1329–1358.

17 Usner, "From Captivity to American Slavery," p. 35; McGowan, "Creation of a Slave Society," p. 35, 138; *MPA*, II, 373; *MPA*, II, 346, 565; Berlin, *Many Thousands Gone*, p. 81, 201. Berlin states that 4,000 of the 6,000 were from Senegambia. Ibid, p. 81. Michael Allen, *Western Rivermen, 1763–1861: Ohio and Mississippi Boatmen and the Myth of the Alligator Horse* (Baton Rouge: Louisiana State University Press, 1990):91; W. Jeffrey Bolster, *Black-Jacks: African American Seamen in the Age of Sail* (Cambridge: Harvard University Press, 1998):45. Hall, *Africans in Colonial Louisiana*, p. 64. For another source discussing African Americans who worked in the maritime trade,

see Peter Linebaugh and Marcus Rediker, *The Many-Headed Hydra: Sailors, Slaves, Commoners, and the Hidden History of the Revolutionary Atlantic* (Boston: Beacon Press, 2000).

18 Hall, *Africans in Colonial Louisiana*, p. 9, 29; Rashauna Johnson, *Slavery's Metropolis: Unfree Labor in New Orleans during the Age of Revolution* (Cambridge: Cambridge University Press, 2016):2.

19 Hall, *Africans in Colonial Louisiana*, p. 20; *MPA*, 2, 573. According to Gwendolyn Midlo Hall, the exchange of survival skills between African slaves and Native Americans tribes—along with other exchanges—contributed to the "multiethnicity and diversity" seen from the beginning of the French presence. See Hall and Charles Henry Rowell, "American Tragedy: New Orleans under Water" *Callaloo*, Vol. 29, No. 4 (2006):1049.

20 Usner, *Indians, Settlers, & Slaves in a Frontier Exchange Economy*, p. 6; John Craig Hammond, "Slavery, Settlement, and the Empire: The Expansion and Growth of Slavery in the Interior of the North-American Continent, 1770-1820" *Journal of the Early Republic*, Vol. 32, No. 2 (Summer 2012):184, 187; Gould, *Fifty Years on the Mississippi*, p. 182. Thomas G. Andrews introduced the term "workscape" as an alternative to "landscape," indicating the perspective of labor. See Andrews, *Killing for Coal: America's Deadliest Labor War* (Cambridge: Harvard University Press, 2010). According to historian John Hebron Moore, by the late 1790s,

> River planters' cotton crops began to bring in immense profits . . . planters of Old Natchez District started the fashion of erecting handsome residence for themselves and superior cabins for their slaves, a custom their successors continued until the Civil War.
>
> See Moore, "Two Cotton Kingdoms" *Agricultural History*, Vol. 60, No. 5 (Autumn, 1986):1–16.

21 Miro to Espeleta, 1 July 1784, Doc. 639, Papeles Procedentes de Cuba, Archivo General de Indias, Seville as quoted in Hall, *Africans in Colonial Louisiana*, p. 212.

22 Numerous texts on marronage distinguish between petit and grand marronage, with one of the best definitions in Terry Welk, "The Archaeology of Maroon Societies in the Americas: Resistance, Cultural Continuity, and Transformation in the African Diaspora" *Historical Archaeology*, Vol. 31, No. 2 (1997):81–92; Richard Price, ed. *Maroon Societies: Rebel Slave Communities in the Americas* (Baltimore: Johns Hopkins University Press, 3rd ed., 1996):149; Sylvia Diouf, *Slavery's Exiles: The Story of the American Maroons* (New York: New York University Press, 2014):4. Unlike many accounts of maroon communities, Diouf discusses the everyday lives of the maroons, drawing upon primary sources such as administrative records and court cases.

23 Din, "Cimarrones" and the San Malo Band in Spanish Louisiana," p. 27; McGowan, "Creation of a Slave Society," p. 227; Diouf, *Slavery's Exiles*, p. 10. Although her book did not focus exclusively on marronage, Sylvia Frey, in her study of the early nineteenth century, viewed "marronage in the New World . . . to have a common African heritage" and not an "innovation in resistance techniques." See Frey, *Water from the Rock: Black Resistance in a Revolutionary Age* (Princeton: Princeton University Press, 1991):52–53.

24 Scholarship on the American Revolution and its impact on African American communities is rich but few examine the role of the Revolution along the Gulf Coast. An exception is Kathleen DuVal, *Independence Lost: Lives on the Edge of the American Revolution* (New York: Random House, 2016).

25 An excellent resource for the Pointe Coupee Conspiracy is "Louisiana Slave Conpiracies" accessed at https://lsc.berkeley.edu/1795. Shannon Lee Dawdy has written an excellent article on Louis Congo, the black executioner of those found guilty in the Point Coupee Conspiracy. In her analysis of the conspiracy and Congo's role, she observed "increasingly "savage" forms of corporal punishment inflicted on black bodies

over time in colonial Louisiana" and how this occurs when the "interests of the state and labor regime intersect." See Dawdy, "The Burden of Louis Congo and he Evolution of Savagery in Louisiana" *Discipline and the Other Body: Correction, Corporeality, Colonialism*, ed. Steven Pierce and Anupama Rao (Durham: Duke University Press, 2006):61–90.

26 George Cable, "Creole Slave Songs" *Century Magazine* (1886) accessed at https://babel.hathitrust.org/cgi/pt?id=nc01.ark:/13960/t8tb3t099&view=1up&seq=9

27 Berlin, *Many Thousands Gone*, p. 201; Hammond, "Slavery, Settlement, and Empire," p. 184; Hall, *Africans in Colonial Louisiana*, p. 9, 120.

2

THE ANTEBELLUM ERA

A River of Contradictions

In 1849, Henry Bibb wrote,

> I have stood upon the lofty banks of the river Ohio, gazing upon the splendid steamboats, wafted with all their magnificence up and down the river, and I thought of the fishes of the water, the fowls of the air, the wild beasts of the forest, all appeared to be free, to go just where they pleased and I am an unhappy slave.

Bibb's prose revealed a riverine environment whose beauty is even more jarring when recognized from an enslaved person's perspective. The beauty of the Ohio River (one of the major tributaries of the Mississippi), the aesthetics of steamboat travel, and all the species that the environment supported are not lost on Bibb. But the incongruence of this imposing landscape and the freedom enjoyed by "fish, fowl and beast" uncovered an environment full of contradictions for him. The inconsistency of a river that indulged the senses and empowered nonhuman species while denying that same freedom to enslaved people was exposed in Bibb's *Narrative* as he related his life as a slave and later, self-emancipation. But in Bibb's story, the river was also the site of freedom as he compared crossing the Ohio River with Moses leading his people out of Egypt. The path to freedom, however, involved a journey through a treacherous swamp-filled riverine environment filled with its own dangers. Bibb's story is riddled with these contradictions.[1]

Henry Bibb was born into slavery in 1815 in Shelby County, Kentucky. His mother, Mildred Jackson, was enslaved and his father was rumored to be a state legislator. Bibb's life in slavery was marked by numerous attempts to escape—prompted by abuse and mistreatment—in which he was initially successful, only

DOI: 10.4324/9781315617077-3

to be captured and forced back into slavery. In his first flight from slavery, his goal was to reach Canada. The opportunity for flight came when he arranged with his "keeper" to look for work in the winter of 1837 and left for the nearby Ohio River. At the time, Bibb and now, his wife, Malinda, and daughter, Frances, all were enslaved at a plantation near Bedford, Kentucky, a short distance from the Ohio River. Ostensibly, Bibb was to work at a slaughterhouse by the river where the wages were high. Instead, Bibb planned on locating a steamer going to Cincinnati. In recounting the first part of his journey, he recalled hearing "the welcome sound of a Steamboat coming up the river Ohio, which was soon to waft me beyond the limits of the human slave markets of Kentucky." Undetected, Bibb reached Cincinnati, where he did not stay but found work in the nearby village of Perrysburg. By spring, Bibb took his earnings and returned by steamboat to his wife and daughter, with plans for their escape north.[2]

By June 1838, Bibb reunited with his family as they hoped to escape north to Cincinnati. The plan was for Bibb to leave first with his family joining him later. Unfortunately, although Bibb reached Cincinnati, he was betrayed and forced to return by steamboat to the slaveholding South. Again, Bibb reflected on the ambiguity of his situation as he contrasted the beauty and solace of the river and its surroundings with the brutality of slavery. On his trip downstream, he observed:

> I could see no possible way of escape. Yet, while I was permitted to gaze on the beauties of nature, on free soil, as I passed down the river, things looked to me uncommonly pleasant: The green trees and wild flowers of the forest; the ripening harvest fields waving with the gentle breezes of Heaven; and the honest farmers tilling their soil and living by their own toil. These things seem to light upon my vision with a peculiar charm. I was conscious of what must be my fate; a wretched victim for Slavery without limit; to be sold like an ox, into hopeless bondage, and to be worked under the flesh devouring lash during life, without wages.[3]

But Bibb's fortunes did not end here as he escaped a second time and returned to Perrysburg with the hope that his family would meet him. When they did not arrive in Perrysburg, Bibb returned to Bedford to ensure his family's flight to Canada. Bibb was betrayed a second time, subsequently imprisoned, and sold along with his family at the Louisville slave market. A slave trader bought Bibb, Malinda, and their daughter, Frances, with the intent to sell them at the New Orleans slave market.

At this juncture, the Mississippi River and its tributaries played another role in the lives of the Bibb family. The inhumane journey downriver began when they were taken in coffles to the Ohio River and later to the Mississippi River to be sold at New Orleans. Coffles involved placing handcuffs on the wrists of the enslaved and then linking them two by two. Imprisoned by the handcuffs and driven downstream by "soul drivers," sleep was difficult, especially when everyone knew their destination. The trip, for Bibb and his family, was a long one, taking six weeks to reach New Orleans, in part due to low water on the Ohio River. The same river that invited escape was now the means to violence and oppression, continuing with their arrival in New Orleans. On reaching New Orleans, Bibb and family were prepared for sale at a slave trader's yard near St. Joseph Street, a few blocks from the Mississippi River. Bibb wrote of the degrading, painful treatment that African Americans received at the slave market as slave traders sought the best price for them. For the Bibbs', they were sold as a unit to a cotton planter with property 50 miles upriver from the mouth of the Red River (a tributary of the Mississippi River in the nineteenth century). Once on the plantation, Bibb recorded the harsh treatment they received as they worked for one of the "most cruel overseers to be found in that section of country." Again, the family attempted to escape, this time starting out in the borderland Red River swamps where Bibb described a landscape marked by "buzzing insects and wild beasts of the forest." When encountering the forbidding, life-threatening landscape with little to sustain them, Bibb reflected,

> What would induce me to take my family and go into the Red river swamps of Louisiana among the snakes and alligators, with all the liabilities of being destroyed by them, hunted down with blood hounds, or lay myself liable to be shot down like the wild beasts of the forest?

Their bid for freedom was in vain, as they were caught and brought back to Whitfield's plantation where Bibb was punished so severely that he could not return to work for several days.[4]

Enduring continued mistreatment, Bibb fled again, only to be caught and sold without his family. Finally, he ended up enslaved to an Indian slave trader, where he successfully escaped, eventually finding his way back to Jefferson City, where through a clever disguise, he boarded a steamboat traveling to St. Louis. From St. Louis, he managed to catch a steamer headed to Cincinnati. Freedom was realized as Bibb recalled the journey north.

> [A] swift-running steamer started that afternoon on her voyage, which soon wafted my body beyond the tyrannical limits of chattel slavery. When the boat struck the mouth of the river Ohio, and I had once more the pleasure of looking on that lovely stream, my heard leaped for joy at the glorious prospect that I should again be free. Every revolution of the mighty steam-engine seemed to bring me nearer and nearer the "promised land."

Bibb remained free, and except for one more unsuccessful attempt to reunite with his family, spent the rest of his years working with abolitionists and publicizing the evils of slavery.[5]

Bibb's *Narrative* encapsulated many of the associations that African Americans possessed of the Mississippi River and its environs. In a sophisticated understanding of the incongruities that slavery wrought, he acknowledged the aesthetics of the river while also recognizing that the river served as his conduit into slavery and freedom. His reference to being without a "Moses" to help in his initial escape and later reaching "the promised land" paid homage to the spirituality of the river and the act of "crossing over" to freedom. But the linkages do not stop here as Bibb's account of the Red River swamp uncovered a harsh riverine environment that simultaneously offered freedom. Bibb's experiences—in all their dimensions—were not unique as other slave narratives shared Bibb's perspective; they contributed to a collective memory of the Mississippi, a shared vocabulary, captured in song, memoirs, folktales, and prose. These collective remembrances expand upon the experiences of African Americans during the colonial era. Adding to the narrative, however, are new developments such as the arrival of steam and more aggressive levee construction, coalescing into a riverine history that not only informs identity but also begs for an accounting.

The antebellum period, which began after the War of 1812 and ended with the advent of the Civil War in 1861, ushered in a period characterized by the exponential growth of slavery in the Deep South, particularly in the Mississippi River Valley. Guided by an emergent nationalism, articulated in popular beliefs, such as Manifest Destiny—a historical undertaking whereby white Americans touted the westward movement as sanctioned by a Christian god—Americans viewed the Mississippi River as a touchstone for U.S. development. The river not only served as an internal waterway fostering explosive economic growth, but it also symbolized the gateway to the West, the frontier. Rich in imagery, the idea of the frontier, embellished with notions of Manifest Destiny, prompted a cultural outpouring in the arts, offering Americans an exceptionalism bolstering a budding nationalism. With the Louisiana Purchase in 1803, followed by the explorations of Lewis and Clark from 1803 to 1805, the United States gained an empire that encompassed the Mississippi River Basin, full of natural wonders, abundant game, and the promise of land. (Of course, the land was already occupied by numerous Native American tribes, a fact overlooked by the federal government and subsequent settlers.) The Mississippi River, as conveyer of goods, national icon, and most important, conduit for the sale of enslaved people, was key to the promising future. By 1860, the value of the Louisiana Purchase was indisputable as more millionaires per capita lived in the Lower Mississippi River Valley than any other region in the United States. The majority earned their income through the production and marketing of cotton and sugar (Figure 2.1).[6]

Up until the outbreak of the Civil War in 1861, slavery loomed large in the southern states bordering the Mississippi River, accounting for the region's unprecedented economic growth. The river was critical—not only for the sale of goods but also for the sale of humans—as one historian noted, "the Mississippi River gave slavery a whole new lease on life." During the period from 1800 to 1860, "at least 875,000 American enslaved people were forcibly removed from the Upper South to the Lower South." Some historians placed the number higher and speculated that as many as one million "were sold down the river." The

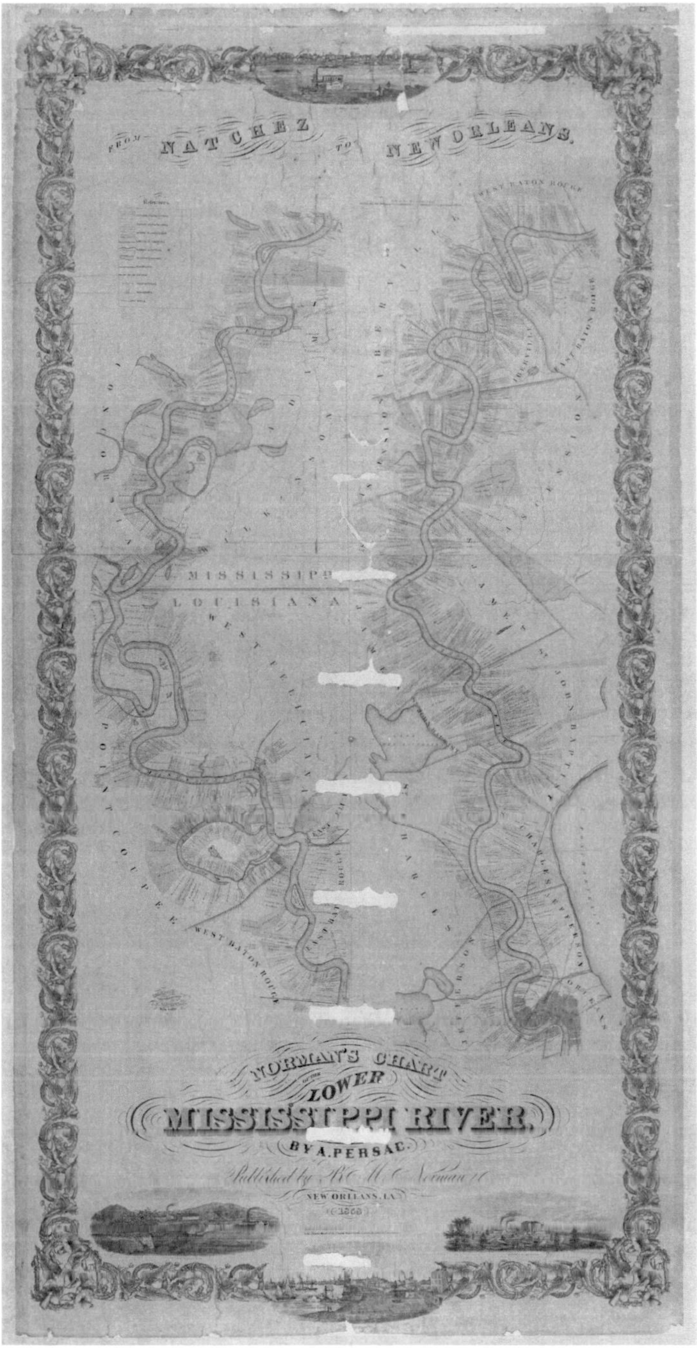

FIGURE 2.1 Norman's Chart of the Lower Mississippi River.

Credit: The Historic New Orleans Collection, L. Kemper and Leila Moore Williams Founders Collection, acc. no. 1947, i–v.

majority of African Americans remained enslaved, providing the labor that made the antebellum market hum. At its height, the United States furnished 80% of the cotton imported by British manufacturers. But enslaved African Americans were forced to labor in other economic sectors as well, including lumber, river commerce, and various trades. By mid-century, expansion beyond the Lower Mississippi Valley had occurred with the promise of more financial gains. Areas such as the Yazoo Delta were tapped for their agricultural wealth, increasing the demand for more enslaved people. Thus, the production of lucrative crops, such as cotton and sugarcane, expanded, with planters always seeking to enhance profits. A lucrative internal slave market ensued as planters in the Upper South states, such as Virginia, realized substantial returns in the sale of enslaved people to markets in the Lower Mississippi Valley. (The interior marketing and sale of enslaved people became so robust during this antebellum period that 30% of the enslaved population could expect to be sold.)[7]

To the enslaved, being sent to the Deep South meant working under some of the worst conditions of slavery as numerous accounts in this chapter illustrate. In one former enslaved person's interview, he remembered how Louisiana plantations had reputations for such brutal conditions that whenever enslaved people were not following the enslavers' rules, they were threatened with removal to New Orleans. In response, the enslaved person would cry for days until the threat passed. But far more draconian measures were taken in reaction to the threat of removal as contemporary stories included one man using an ax to cut off the fingers of one hand in order to avoid being "sold down the river." Another man shot himself and finally, a mother with several children drowned herself in the river rather than being sold downriver. The reactions were well founded as conditions on the plantations of the Lower Mississippi Valley were intolerable, borne out by the frequency of violent acts against the overseers. For enslaved people working on the sugar cane plantations, for example, the hours were long during the "sugar-making season." Although sugar was one of the more profitable crops, Louisiana's climate was not as conducive as that in in the West Indies, where sugar was the main crop, resulting in a shorter time frame. Thus, in Louisiana, the window for cultivating and harvesting was shorter with 16-hour days of strenuous, back-breaking labor for the enslaved. Another facet of being sent to work in the region included a trip down the Mississippi River—in coffles as described by Bibb—followed by the degrading slave auction at river ports such as Natchez and New Orleans. Complementing Bibb's memories were those of countless others. One witness to the trip downriver was Charles Sealsfield, an Austrian-American journalist, who, when traveling down the Ohio River by keelboat in the mid-1820s, recalled:

> Forty slaves who were to be transported to the states of Mississippi and Louisiana, were a sort of deck passengers . . . as long as the weather continued fine, the poor negroes had a tolerable lot, but when afterwards it began to rain, and they continued on deck seven and a half feet broad, and

> forty-two long, without any covering over their heads, or being able to move, our kitchen being likewise upon deck, their situation became truly distressing, and one of the infants died shortly afterwards; another, as I was informed, fell into the Mississippi above Palmyra settlements.[8]

Memories such as these were not unusual, and historians now refer to the trip down the Mississippi to the Deep South as the second middle passage.

With atrocities such as these—laboring on the sugar plantations or the trip downriver—juxtaposed with the immense wealth accumulated by Southern planters, perhaps nowhere is there a greater divide regarding perceptions of the Mississippi River than in the antebellum era. The river for white planters was part of an emergent, wildly successful global market, a means to transport not only goods, but also their labor force. For African Americans, the antebellum river facilitated a new low, the nadir of American slavery. In sharp contrast to a popular imagination where the river represented profit and iconic value, to the enslaved, the river, in this instance, was the conduit into the bowels of slavery where families were separated and oftentimes the harshest abuses occurred. Thus, at the height of a Mississippi River-based commercial empire contributing to vast fortunes for many, who in turn were dependent upon the forced labor of an estimated one million enslaved blacks taken down the river to be sold, the river had an equally compelling connection to most African Americans. The river represented violence and trauma on an unprecedented scale while offering a means of escape. Associating the river as a site of violence and trauma became a mainstay in African American experiences with the river, a constant in the river's history. As Bibb articulated in his *Narrative*, the river embodied extreme contradictions, celebrated and lamented in the cultural venues of song, folklore, and prose.[9]

For African Americans, the violence associated with the river continued throughout the antebellum era and persisted into the twentieth century. Following a trajectory that began in the colonial era with the executions and punishments endured by San Malo and his maroon community, the river again became a site of violence in an unsuccessful slave revolt in 1811. Led by Charles Deslondes, the revolt became known as the Deslondes Revolt. Often downplayed in the literature on slave rebellions, the revolt included more than 500 enslaved people along the German Coast of the Mississippi River. Fueled by the Haitian Rebellion, with many enslaved people possessing tracts of the French *Declaration of the Rights of Man*, the rebels marched along the River Road with the intent to enter the city of New Orleans. The goal was a free African American state along the Mississippi River. Given their numbers and the destruction the rebels were causing in route to New Orleans, federal troops were called in to stop them. Through the combined efforts of federal troops and the state militia, the uprising was quelled with a bloody aftermath. To squash any future rebellions and reassert power, some of the revolting enslaved people were dismembered with body parts put on full view. Others were arrested and sentenced to death, after which they were decapitated with their heads displayed on poles lining the river for more than 60 miles. Although the "Ballad of San Malo" referenced the display of the heads of captured maroons on surrounding levees, it was the Deslondes Revolt where this occurred, prompting one journalist to remark "the levee . . . is ornamented with poles, on which are placed numbers of the heads of the unfortunate wretches." In both instances, the river's history became entwined with not only visions of freedom but also the site of numbing violence—a sharp contrast with the river of a Mark Twain or Frances Trollope.[10]

Juxtaposed with the fortunes of those associated with the Deslondes Revolt was the commercial success enjoyed by those engaged in cotton and sugarcane production, none of which would have been possible without the advent of the steam engine. In the Mississippi Valley soon after the first steam-powered ship voyaged up the river in 1811, steamships overran the riverine landscape. Before 1817, there were only 20 barges annually serving the river from the mouth to the upper Mississippi. But by 1860, there were more than 3,500 boats arriving at the New Orleans levee annually. By the 1840s, steamboats had made it into every inlet of the Mississippi River. Steam-powered boats dramatically shortened the journey between markets. For example, before steam, the best record for travel between New Orleans and Cincinnati was 78 days. By 1817, steamers took 25 days to reach Louisville from New Orleans. To further indicate the scale, these boats represented 16,000 tons of shipping with a capital investment of $17 million. Put another way, in the early 1800s, an estimated five million pounds of cotton were shipped from the Mississippi Valley compared with 200 million pounds in the 1830s. As a result of increased shipping, the cotton empire facilitated the rise of New Orleans as it became one of the busiest ports in the world by the mid-nineteenth century. By 1860, the city had grown almost ten times since 1810, with a population comprising half white and half black, and with a small percentage of free blacks. Accompanying the commercial use of steam transportation was the rise of a burgeoning tourist industry. The river as a national icon gained traction as well-off tourists enjoyed the riverine scenery from ostentatious "floating palaces." Travel literature became popular reading with accounts regarding Mississippi River travel from such cultural icons as Charles Dickens and Francis Trollope. Celebratory tours such as the Grand Excursion of 1854 nurtured the river's place in an emergent national identity, further enhanced with the poetry of Henry Wadsworth Longfellow, 4-mile-long panoramas of the river, or the artwork of Jacob Bingham reminiscent of the age of the flatboatmen. Yet, despite the mass popularity of the river for travel and cultural production, most steamboats supported the commercial enterprise of a growing cotton empire in the American South.[11]

With the rise of steamship travel and commerce, the river's role in African American lives' expanded. Steamships provided jobs for enslaved and free blacks as stewards, porters, deckhands, housekeepers, cooks, and firemen. Certain jobs were more valued than others, as one 1847 account recalled "a great living as a porter." Less desirable jobs included work as roustabouts, loading and unloading goods, although before the Civil War most roustabouts were of Irish or German descent. New tasks emerged with the steam engine such as getting wood, known as "wooding up," to feed the boilers. These jobs were often assigned to free and enslaved blacks. In the words of one traveler aboard a first-class steamer in 1855, 60 men might be charged with gathering 100 cords of wood from the shore. The work was well orchestrated.

> The laborers pursue their calling with the precision of clock-work. Upon the shoulders of each are piled up innumerable sticks of wood, which are thus carried from the land into the capacious bowels of the steamer. The "last loads" are shouldered—the last effort to carry the "largest pile" is indulged in.[12]

Once the "wooding-up" was completed, an equally demanding job was ensuring the boilers stayed full in order to power the boat. This work was physically challenging on two fronts—the strength required and the intense heat from the boiler. Songs were often invoked to capture the rhythms of the work, as one European traveler recalled:

> The immense fire-engines are all on this deck, eight or nine apertures all in a row; they are like yawning fiery throats, and beside each throat stood a negro naked to his middle, who flung in fire-wood. Lerner H. encouraged the negroes to sing; The negro up aloft on the pile of fire-wood began immediately an improvised song in stanzas, and the close which the negroes down below joined in vigorous chorus. . . goes on to remark, "while they, amid their equally fantastic song, keeping time most exquisitely, one piece of fire-wood after another into the yawning fiery gulf.[13]

Another recollection of life on the steamers for enslaved people came from George Forman, who worked on steamships before the Civil War. He recalled "Many boats had negro crews of fireman, and on the Boat backing out from the Levee at St. Louis or New Orleans or Cincinnati, they would gather at the Bow, the leader of the singing standing on the Capstam waving his hat, and singing—He leading and the crew joining in the refrain or chorus, "We are bound for the Crescent City, Refrain: Hooray You Rolling river, Where the Yaller Gals they are so pretter. Refrain: Ah Ah You Shennadoah." . . . And other songs of that kind. "Fire down below" &c." These work songs would also be associated with the call and response one heard in the fields, fostering a collective memory of the river that would be foundational for the next century when the "flood blues" further cemented the relationship between African Americans and the river.[14]

Enslaved people were also "hired out" by enslavers to work on the steamships in a variety of other positions. For example, in one 1850 account from an enslaver in St. Louis, he discussed how his slave, Frederick, was the chief cook on board. In correspondence from another enslaver, he complained that the steamship owner was not paying his slaves' wages of $40 a month. Other types of black labor in river trades can be found with one contemporary's account of a boat being unloaded in New Orleans, where the "negro sorting clerk" was particularly proficient. Still other jobs, mentioned in firsthand accounts, included work as a boat furnisher or in the words of one St. Louis resident who claimed "the river provided him with a handsome living" as a produce trader in the 1840. Whether porter, clerk, or boilerman, these jobs allowed a certain freedom and a greater possibility of escape. For African Americans, where steamboats could be found docked in the major port cities, the opportunities for communication multiplied, creating networks for enslaved and free blacks. The planters, however, were also very aware of the increased risk of escape for those working near the docks or aboard the steamships. As regional newspapers in the Mississippi Valley advertised for the return of runaway slaves, references to a potential escape by way of the river provided further evidence of the river's role as the means to freedom. For example, in the *Louisiana Gazette* dated May 1805, one enslaver offered a reward for "Charles" who "will pretend to be free and try to get employment on board a ship and get a berth to go abroad." In another ad placed in 1829, an enslaved laborer named Briton had run away, dressed as a sailor with hopes to board a ship. Thus, as steamboat travel increased, the number of runaways also grew, particularly along the Lower Mississippi River. Many enslavers saw proximity, whether in port cities such as New Orleans or a riverside plantation, to the river as another opportunity for escape. Their fears were realized as it was common for enslaved people working on riverside plantations to attempt escape. When arguing for the Fugitive Slave Act in 1850, Jefferson Davis, contended that planters with riverside plantations often saw their land devalued because of the perceived ease for a slave's escape. Despite the perceptions and attempts, however, successful escapes for enslaved people were few. As a result, slave catchers filled the jails in New Orleans, Natchez, Baton Rouge, and Vicksburg with runaways, prompting concern by the enslavers about the health of the runaways. Yet, the perception that a river escape was viable remained for both enslaved people and the enslavers casting the river as a means to freedom for the enslaved.[15]

In addition to opportunities for escape, working on steamboats afforded prospects for illegal activities. In a well-known case in 1841, the river was again a site of resistance and violence as four black men were publicly executed in St. Louis. Three of the men—Amos Warrick, James Seward, and Charles Brown—were free whereas the fourth, Madison Henderson, was enslaved. All four worked on steamships in jobs ranging from cooks, stewards, waiters, and boilermen to deckhands. Their crimes, which were not disputed, included theft, murder, and arson. Upon capture, they were tried, found guilty, and punished by hanging on Duncan Island. After the public executions, they were decapitated with their heads placed in a St. Louis storefront. These men, who had committed other crimes as well, called their activities, "rascality," representing another form of resistance along the Mississippi River. Once caught, however, their trials and executions offered an opportunity for the slaveholding South to reinforce its control over this racialized society. Southerners were always alert to the potential of escape by means of the river, with African Americans working on steamers heightening the concern. The executions became a showpiece, a theatrical display of white dominance as tickets were sold for people to board steamships and witness the hangings. One account claimed a crowd of 20,000 to 35,000 purchased tickets to watch, whereas another contended a more modest audience of more than 20,000 purchased tickets. For African Americans, the event contributed to a growing collective memory where the river and its environs would be the site of violence—whether in the treatment of those revolting or those committing acts of "rascality." A trajectory of river-associated violence punctuated subsequent decades as the post–Civil War era witnessed lynchings in Delta lands, another dimension of the Mississippi River's history (Figure 2.2).[16]

FIGURE 2.2 Advertisement to Watch the Hangings of Warrick, Seward, Brown, and Henderson.

Credit: Courtesy of Missouri Historical Society Library and Research Center.

Closely tied to the rise of commercial steam travel on the river was the need for ongoing construction and maintenance of levees. From the onset of European settlement in the Lower Mississippi Valley, settlers sought to control the river through levees. Beginning with colonial correspondence from the first French governor of Louisiana Territory, Jean-Baptiste Le Moyne de Bienville, the need to construct levees to protect the city of New Orleans and nearby plantations was always referenced. The nineteenth century continued the pattern. In numerous accounts, slaveholders recorded the work performed on the levees; thus, by 1861, there were more than 2,000 miles of levees lining the Mississippi River. These levees, in the states of Arkansas, Louisiana, and Mississippi, averaged a height ranging from 8 to 10 feet with a width from 50 to 75 feet. The cost of building the levees was significant and by 1860, the Louisiana State Engineer estimated that the levees in Louisiana alone had cost $12,500,000.00. Enslaved labor accounted for most of the work of protecting the plantations and river cities, particularly New Orleans. For example, in the records of one plantation overseer, he recorded almost daily the number of enslaved people assigned to levee work, with entries such as one dated January 19, 1861, in which he noted "22 men at river on back levee." But the work was not limited to men as another day, he recorded: "20 women with hand barrows making back levee." Levees remained a constant in the African American imaginaries of the Mississippi River, prompted by several historical incidents. In addition to the association connected with labor, levees were also linked with violence as the discussion of the Deslondes Revolt revealed. The *Dirge of San Malo* first cultivated the imagery of violence with reference to the decapitated heads of San Malo's community displayed on the levees. Although historically inaccurate, San Malo's ballad preserved the imagery and the violent association with levees. After the Civil War, most of the levee system was destroyed, leaving the U.S. Army Corps of Engineers faced with the herculean task of rebuilding. Levee camps were set up, with African Americans comprising the majority of the workforce. Despite the end of slavery, new tales emerged regarding the barbarity of the camps—rivaling earlier abuses of slavery.[17]

As the institution of slavery grew in the Deep South, facilitated by the arrival of steam and a variety of short-staple cotton, thereby fueling a more prosperous, affluent Southern society, representations of the river crystallized into competing narratives. For African Americans, new linkages evolved from experiences with steamboat travel—whether through work, escape, or the proliferation of levees—whereas earlier representations remained. Old or new, differing narratives emerged for African Americans and whites, revealing a multidimensional Mississippi River with a complexity that defied simplistic, dualistic attempts to define the waterway. Yet, constants remained as African Americans experienced the river through labor, refuge, and resistance. For those enslaved, the river might be the conveyer of the worst abuses of slavery. In contrast, to the planter—fast emerging as one of the nation's wealthiest classes—the river represented a means to wealth. But there were other representations as the river also signified freedom for African Americans; a means to escape north. The Ohio River, one of the Mississippi's largest tributaries, particularly became associated with freedom as crossing the Ohio meant crossing over into freedom. The associations do not end here as the backwaters of the river, supporting the swamps and bayous, also offered an alternative geography for those enslaved. For African Americans, the colonial era narrative of the river with all its contradictions remained although accelerated and expanded through the advent of steam and the increased number of sugarcane and cotton plantations.

Yet unlike the colonial era, the antebellum period offers more firsthand accounts of slavery as self-emancipated slaves shared their experiences, exposing the horrific institution of slavery and giving voice to another Mississippi River. Within the autobiographies, former enslaved, forced to work in the Deep South, revealed the ongoing narrative of the Mississippi River through the lens of the African American community. Through their histories, complemented by song and folklore, the recurring themes of oppression, liberation, refuge, and resistance characterized the intersection of African Americans with the Mississippi River. Building upon the associations forged during the colonial era, the riverine environment evolved into a racialized space as swamps developed not only into more than a refuge for runaways but also spaces separate from the planters' oversight. These spaces might offer a site for conspiratorial meetings, religious gatherings, or temporary respites from slavery. Regardless of use, the surrounding swamps allowed for a negotiation with the environment, free from oversight. But firsthand accounts also uncovered other racialized spaces where blacks were empowered to live and work outside the white hierarchy. The following accounts, supported by newspaper advertisements and contemporary scholarship, illustrate an increasingly multidimensional understanding of the river, tempered by race and class. Throughout the nineteenth century, these considerations of the river solidified and by the twentieth century, African Americans integrated a history of the river, as seen by song, prose, and oral history, while converging at times with the dominant narrative, distinct in its associations and representations.[18]

In all the firsthand accounts, labor underlined the African American experience with the river documented by well-known images of enslaved laborers working on the levees, plantations, or the numerous steamboats. Yet often overlooked in histories focused upon the exploding economy of Southern planters and their exploitation of black labor is the internal economy that black labor cultivated from the largesse of the Mississippi River. Beginning in the colonial era with its "frontier exchange" economy, by the antebellum era African Americans maintained an economy that relied upon harvesting the surrounding environment. In many cases, the economic exchanges were covert. For example, despite complaints from planters that enslaved people were stealing and selling their chickens to river traders, the practice continued throughout the 1800s. River traders, known as "chicken thieves" to the planters, traveled throughout the Mississippi River's waterways and back streams, bartering goods with the region's laborers. According to one 1850s' contemporary, the itinerant peddlers were more troublesome along the Mississippi "Coast" than any other. Arriving at night, one former bondsperson recalled that enslaved people would "git down to de ped'lers on de riber at nite tuh buy stuff." In one particularly wealthy region in southern Louisiana where sugarcane was cultivated, records indicated that enslaved people tended gardens, raised poultry, and hunted, selling the fruits of their labor. There was also a market for Spanish moss as well as firewood that was sold to the steamboats. In the words of one historian, these internal economies "shaped patterns of slave life" and resulted in the enslaved person's ability to supplement diets or buy additional goods, such as clothing. The river offered sustenance whether through transport or goods dating back to the years of the French settlement.[19]

One of the most enduring internal industries realized from the riverine environment was the selling of Spanish moss. Used for stuffing summer mattresses, black laborers harvested Spanish moss from the cypress trees that dominated the swamps. Sometimes, these internal economies had the blessing of the enslavers as enslaved people realized significant profits. In one firsthand account, an enslaved West African man, nicknamed "The Prince," and formally known as Ibrahima, was allowed by his enslaver to collect Spanish moss and sell it at the weekend market in Natchez. Ibrahima also sold produce from his garden. Attending the market not only provided Ibrahima with a small amount of cash but also allowed for contact with other enslaved Africans from whom he might hear news of his home. Eventually regaining his freedom in 1828, Ibrahima related his fascinating life story as he was the son of a king, and as such was well educated. One weekend while at the Natchez market, someone from home recognized Ibrahima, calling out to him by his proper name "Abdul Rahahman [Ibrahima]" and kneeling to the ground with his face turned to the earth. He brought news to Ibrahima of his father's death and the succession of his kingship. How often the market served as a network—granted that Ibrahima's experience was unusual—for enslaved African Americans is unknown, but evidence exists that these internal economies not only furnished income but also offered opportunities to meet and exchange information.[20]

In addition to gathering Spanish moss, black laborers worked in the swamps, cutting down the valued cypress trees as the early nineteenth-century popular historian Timothy Flint described. After finding the best trees

> The negroes surround the tree in periogues, and thus get at the trunk above the huge and hard buttress, and fell it with comparative ease. They cut of [sic] the straight shaft, as suits their purpose, and float it to a raft, or the nearest high grounds.

In addition to supplying lumber, and in some instances enslaved people acting as independent contractors, the swamps served other purposes, as they did in the colonial era. Many of the slave narratives included passages about escaping through the swamps or hiding in the swamps on their way to freedom. When John P. Parker recalled his night in the swamp, he leads the reader through his first steps in this soggy ground with his feet sinking into the ground until he finally saw what he called "four forest friends," which were trees intertwined, offering a canopy over the marshy ground. Parker slept in one of the trees' hollows and the next morning after finding berries to sustain him made his way to a nearby road. On his way out of the swamp, he commented upon the "dreary" swamp terrain and its "absence of life." In this instance, the views of African Americans and whites converged as Flint described the environment of the valued cypress trees,

> The cypress loves the deepest, most gloomy, inaccessible and inundated swamps; and south of 33 degrees, is generally found covered with the sable festoons of long moss, hanging, as it seems, a shroud of mourning wreaths almost to the ground.[21]

Reinforcing these grim associations were those of Solomon Northup, who when attempting to escape found himself in the "Great Pacoudrie Swamp." Northup expounded on his time in the swamp but unlike Parker found it full of life, albeit "wild beasts" that included snakes, tigers, bears, alligators, and wildcats. The wildlife he encountered, which included scores of quacking ducks, frightened Northup as he plunged deeper into the swamp. Yet, Northup appreciated the life-giving properties of the swamp as he remarked "Even in the heart of that dismal swamp, God had provided a refuge and a dwelling place for millions of living things." In other slave testimonies, escapes to the swamps occurred for either short- or long-term absences. Toward the end of the Civil War, Octave Johnson recounted running away to the "woods," and living in a cypress swamp for one and a half years. During this time, he said the group he was with grew to 30 people. In recalling their survival strategies, he related how they "slept on logs and burned cypress leaves to make a smoke and keep away mosquitoes."[22]

Throughout the antebellum period, the Mississippi River swamp and the surrounding wilderness were alternate geographies, competing landscapes that served multiple purposes. First, these uninhabited grounds provided refuge, whether for maroon communities who might survive for years as they hunted and harvested the swamp's rich reservoir of natural resources or as temporary shelters in route to the river. Second, slave narratives credited the swamps as sites for daytime shelter or places that supported the internal economy of selling Spanish moss or harvesting cypress trees. Third, in numerous records, enslaved people referenced meeting in the woods or swamps for religious services or other covert activities. In the words of Peter Randolph, a former slave from Virginia, swamps were "places to meet with and talk with God." The following song reflected the dual meanings of the wilderness—physical safety and a spiritual haven.

> I sought my Lord in de wilderness, in de wilderness
> I sought my Lord in de wilderness, For I'm a going home
> I found free grace in the wilderness.

Songs such as this revealed another facet of African American life in which the wilderness, or in this case, Mississippi River bayou, was not only liberating but also empowering.[23]

For enslaved people living by the bayous and swamps, they recognized the benefits that the swamp provided in their daily existence. During the Lincoln administration, one member of his cabinet charged with reporting on the lives of soon-to-be emancipated slaves, when visiting southern Louisiana, remarked upon the number of escapes to the nearby swamps. The report concluded:

> The swamps were never free of negroes. They constituted a species of asylum, and that fact had its effect upon the character of the negro, and upon the working of the system. As a general thing the negro became more self-reliant, and the master more.[24]

Thus, outside the white hierarchy of domesticated landscapes, the marginalized swamps were an arena where enslaved people had control. Concomitant with the use of swamps for meetings or escape, the swamps lent another facet to African American life in the Mississippi riverine environment. As watercourses ran through the swamps, distinguishing the hinterland behind the riverfront plantations, African Americans acquired a geographical literacy about their landscapes—evidenced by the "self-reliant" characteristic referenced earlier—empowering the community. At the forefront of this geographical knowledge was a sense of mobility that characterized life in the riverine environment. Attempted escapes, river work, and later migrations north exemplified the mobility that surrounded them as they read the landscape. The association of the river with mobility for African Americans persisted throughout the twentieth century. In the words of one scholar, studying the Natchez district, "The fields and the great house were places of work and struggle. The wild places were good for worship and running away." Far from the watchful eyes of planter and overseer, the hinterland afforded autonomy as well as a site for community and gathering. The swamps, along with the numerous waterways that marked Delta plantations, became in the words of one scholar "an alternative territorial system."

Thus, exercising control over physical spaces outside the planters' immediate milieu contributed to an autonomy within the culture of enslaved African Americans. Mastery of these spaces resulted in avenues of illicit trade, clandestine meetings, and escape, an arena far from the planter's oversight. These marginalized lands, to borrow a term, became "black spaces," remembered and celebrated through song, informing a cultural identity—in this instance, derived from the Mississippi River largesse—independent of the enslavers. Exercising control over swamplands—whether through harvesting resources for an internal economy, finding refuge, or serving as spiritual haven—fostered a collective memory, where the riverine environment was more than a site of violence and trauma. Instead, the river and its environs emerged as a site of nurturing and resistance as enslaved people re-envisioned and refashioned the landscape to support their physical and spiritual needs. The duality that Bibb recognized was reinforced in the swamps and wilderness as sites offering both solace and violence.[25]

Yet the swamp was not the only landscape associated with the spiritual world of African Americans. The river also played a religious role, becoming internalized as a site with spiritual properties evidenced by spirituals such as one recalled by Ellen King, a former slave from Mississippi:

> Down by the river side,
> Jesus will talk and walk,
> Ain't going to study the world no more,
> Ain't going to study the world no more,
> For down by the river side,
> Jesus will talk and walk.

Although this spiritual was sung in other geographical areas across the United States and not written with the Mississippi River specifically in mind, the historian Jon Sensbach contended: "the song distills the notion of the constant, intertwined presence of water and Jesus, and the Mississippi was the omnipresent water in the lives of thousands."[26] Another spiritual celebrating the role of the river and heard throughout the South was the "Ship of Zion," with the following lyrics:

> I was standing by the banks of a river
> Looking out over lifes troubled seas
> When I saw an old ship that was sailing
> Is that the old ship of Zion I see
>
> It's hull was bent and battered
> From the storms of life I could see
> Waves were rough but that old ship kept sailing
> Is that the old ship of Zion I see
>
> At the stern of the ship was the captain
> I could hear as he called out my name
> Get on board It's the old ship of Zion
> It will never pass this way again
>
> As I step on board I'll be leaving
> All my troubles and trials behind
> I'll be safe with Jesus the captain
> Sailing out on the old ship of Zion.[27]

Further testifying to the symbolic role of the "Ship of Zion," according to one historian, one former slave recalled her mother who when first seeing a steamboat believed that she was witnessing "the "Old Ship of Zion" come to take her to heaven." Finally, in another song, "Sinner Won't Die No More," the river is directly referenced:

> O de Lamb been down here an' died . . . .
> Sinner won't die no mo'
> I wonder what bright angels . . . ., De robes all read now
> O see dem ships come a-sailing, sailing, sailing,
> O see dem ships come a-sailing,
> De robes all ready now.

Of all the spirituals related to the river, perhaps the most well known was "Crossing Over Jordan," invoking memories of crossing the Ohio River or Mississippi River to freedom, giving new meaning to the following lyrics:

> Deep River, my home is over Jordan.
> O don't you want to go to that Gospel Feast,
> That Promised Land where all is Peace?
> Deep River, I want to cross over into that Camp Ground.[28]

Another facet of the river's spiritual properties was revealed in the ritual of baptism, particularly the practice of immersion popularized by the Baptists. One of the groundbreaking works considering African American religious traditions contended that the practice of immersion, a staple in the Baptist church, demonstrated the persistence of West African-based religious influences, specifically the cult of the water spirits. In accepting Herskovits arguments, the predisposition to understanding rivers and water as purifiers through baptism and passages to the next world, through the act of "crossing over," contributed to furthering the collective memory of the river as refuge. One of the songs celebrating baptism included the "Baptizing Hymn" with the following lyrics:

> Freely, go marching along
> Down into the water
> Freely, go marching along
> Like Zion's songs and daughters
>
> Ev'ry time I look up the House of God,
> The Angels cry out Glory:
> Glory be to my God who lives on high!
> To save a soul from danger.[29]

In addition to riverside baptisms, enslaved people utilized the surrounding woods and swamps to conduct prayer meetings. As many enslavers banned religious services, enslaved people became very adept at avoiding detection; as one former slave from Louisiana recalled, they

> would form a circle on their knees around the speaker who was also on his knees. He would bend forward and speak into or over a vessel of water to drown the sound. If anyone became animated and cried out, the others would quickly stop the noise by placing their hands over the offender's mouth.

African Americans also developed a code to inform each other when a meeting was to be held. For example, the song "Steal Away to Jesus" signaled that a prayer meeting would be held that evening. Songs associated with prayer gatherings occurring in the bayous and swamps, along with the call and response field work songs, were the nucleus of an African American culture, fulfilling a hierarchy of needs. By the twentieth century, this culture of song would produce the blues and other musical genres (Figure 2.3).[30]

But mastering the landscape of the dreary, forbidding swamps invited other problems for African Americans. Publicizing their knowledge about the swamplands and wilderness bordering the plantations would not be without costs. Well-known African Americans such as Frederick Douglass realized that if blacks appeared too adept in the inhospitable borderlands, filled with "wild beasts" and other threats, their abilities would be dismissed as proof of savagery. As African Americans sought to end slavery and appear ready for citizenship, their abilities in the swamps needed to be tempered. A fine line presented itself:

FIGURE 2.3 A Plantation Burial.

Credit: The Historic New Orleans Collection, L. Kemper and Leila Moore Williams Founders Collection, acc. no. 1960.46.

pride in the survival skills necessary for escape through treacherous terrain while still demonstrating their humanity and equality. Still planters relied upon the enslaved people's wilderness skills and often depended upon them for hunting game. In addition to hunting for the enslavers, many enslaved people also hunted for their families who relied upon the extra nourishment that wild meat provided. In his narrative, Solomon Northup several times referenced his abilities as a hunter—catching and eating "roasted possum" or setting up a fish trap. Northup, like others, realized the importance of acquiring extra food sources, given the substandard diet furnished by the slaveholders. Another African American enslaved along the Mississippi River, John P. Parker, boasted of his fishing skills and how the other enslaved people always wanted to accompany him. In nearby Alabama, Julie Scott reminisced how their diets were supplemented with wild game from racoons, rabbits, bears, and even turtles. But in another interview, Sally Murphy contradicted Scott and claimed that the men in her family only hunted possum and wild turkey and did not bother with rabbits or "waste time on fishing." Other types of game that enslaved people hunted and trapped included wild cats, squirrels, and "varmints." Ecological knowledge of the environment extended to herbs and plants used for medicinal and healing purposes. As early as DuPratz's time in French colonial Louisiana, African American narratives testified to this knowledge. Thus, despite the potential consequences from mastering an environment shunned in its pristine setting by enslavers, African Americans derived many benefits from their ability to navigate the swamplands. In addition to the practical advantages—a source for game, roots, and herbs—successfully navigating this environment empowered African Americans as they prospered and thrived in a setting where whites did not. This fact alone dismantled white predominance and while the swamps offered sites of resistance, the landscapes also inspired and emboldened a collective memory proffering African American mastery over their surroundings.[31]

One other population inhabited the wilderness and swamps bordering the Mississippi River—the maroons. Unlike the colonial era where the well-known daring of San Malo and his community became legend, marronage in the nineteenth century was without such heroic tales. Yet, accounts of maroon communities can be found in contemporary newspapers. In 1827, *The New York Post,* for example, included a story about a black woman who returned to her "master" after spending 16 years in a maroon community with 60 other people, located 8 miles outside of New Orleans. Other accounts, such as one in 1836 that mentioned "runaway negroes in the cypress swamps" near New Orleans, were not unusual. Similar to colonial era communities, the maroons harvested their surroundings, incorporating a hybrid economy that relied upon bartering with enslaved people from nearby plantations. For example, in Octave Johnson's account, he discussed how his maroon community stole cattle and traded the butchered meat with nearby enslaved people in return for cornmeal. Other accounts included one of a man sitting on top of a cypress tree for three years and avoiding capture. Similar to the colonial period, marronage in the antebellum era testified to the resistance of enslaved people and the geographical literacy that facilitated it.[32]

For most African Americans, however, a more common experience was working on the river, particularly after the arrival of steamboats, when 20,000 free and enslaved blacks labored on western rivers. Numerous firsthand accounts reflect on lives working on the Mississippi. For example, in a 1937 interview, George Burns observed: "The life of a riverman is a good life and interesting things happen on the river." Burns' biography included harrowing, near-death experiences as his childhood, when still enslaved, was marked by the loss of his toes from freezing night temperatures. Early in his life, he was indentured to a flatboat captain where he worked several jobs beginning with dishwashing. Limited by his physical condition, for the rest of his life, he labored on boats, and in his words, "I know steam boats from wood box to stern wheel." Having lived a full life on the river, he shared many memories. For example, he recalled the bustle of the Louisville and even busier, New Orleans' market. But he also remembered the "slave pens" in New Orleans and the hideous practice of auctioning off slaves. Burns' time on the river featured a near-death experience as one of the steamers in which he worked wrecked, leaving him struggling to survive in the water before finally being rescued. With pride, he discussed working on the *Eclipse*—the longest steamboat constructed in the West—and the speed at which it moved. Juxtaposed with this memory, however, was another one recalling the law that required him to carry a pass at all times so "we could save our skins if we were caught off the boats." Burns' recollections mirrored those of other African Americans during the steam age. Aware of the benefits of river work, Burns still did not overlook the drawbacks. While he contended that working on the river offered a good life, Burns did not diminish what laboring on the river, for a black man, entailed. His story revealed not only the worst abuses of slavery, facilitated by the river, but also the opportunities that the river presented. In his description of the activity surrounding the river whether through markets at New Orleans or people lining the riverbanks and greeting an incoming steamboat, he conveyed an environment that hummed with activity, much like the one described by James Thomas, a freedman and successful businessman. One of a minority of free black men working on the Mississippi in the 1850s, Thomas described the St. Louis wharf "as a vast warehouse without a roof." These narratives of the river converged with popular imagery of the antebellum river. Another key theme that emerged from Burns' account and others is the mobility that river work provided, unlike other types of labor.[33]

River piloting, although a less common profession for African Americans, was recorded, illustrating an awareness and mastery of the Mississippi. Knowledge of the river's defiant and at times, life-threatening, channels was often the subject of nineteenth-century river pilots, with the most notable being Mark Twain. The exploits of steamboat captains were celebrated in numerous memoirs such as George Merrick's *Old Times on the Mississippi*, staples of popular culture. Less well known but equally impressive was the account of Josiah Henson, born in

1789 in Maryland and who later moved to Kentucky before escaping slavery in 1830. Before his self-emancipation, Henson worked on the river taking goods by flatboat to New Orleans. In his autobiography, Henson observed that since he was the "only Negro" on the boat, he was expected to do more than the others. As a result, Henson claimed, "I learnt the art of steering and managing the boat far better than the rest." In the familiar prose of river pilots, Henson wrote: "I watched the maneuvers necessary to shoot by a sawyer, to land on a bank, or avoid a snag, or a steamboat, in the rapid current of the Mississippi, till I could do it as well as the captain." Henson's skills were tested when the captain was blinded by the sun's rays reflecting off the water and Henson became "master of the boat" for the rest of the journey to New Orleans.

His experiences, however, were not isolated ones as Allen Sidney, who prior to the Civil War learned how to operate a steamship running from Pittsburgh to Memphis. In a post–Civil War interview, Sidney said that he continued operating boats for the next seven years. In his role as river pilot, however, Sidney also possessed entrepreneurial skills; thus, by the time he was freed after the Civil War, he had saved $150 dollars in gold, as a result of taking chickens down to New Orleans and selling them. Sidney's experience reinforces the existence of the ever-present internal economy that characterized the period. Still another exceptional river pilot was Simon Gray, who lived and worked in the Natchez area during the late antebellum period. Enslaved by a sawmill operator, Gray was entrusted to pilot flatboats and rafts filled with cargoes of cypress lumber downstream to New Orleans. At times, he even delivered lumber to riverfront planters on the German Coast. With the autonomy and responsibility Gray possessed, he accumulated significant wealth, allowing him to indulge in such practices as visiting an area hot springs for his health. The visit cost Gray $1,000.[34]

When working on the river, songs were sung, recalled by those traveling on the Mississippi. For example, when rowing, work songs captured the rhythm of the water's constant motion, prompting an intimacy with the southern byway, a knowledge of the river's mechanics through work. In George Washington Cable's nineteenth-century work on Creole slave songs, he included a section on songs of woods and waters. Recognizing the ever-present waterway or bayou front in the Lower Mississippi Valley, Cable recounted the "Rower's Song," which in his words "the throb of the song measured the sweep of the oars." A sampling of the song includes the following:

> Sing, lads; our master bids us sing.
> For master cry out loud and strong.
> The water with the long oar strike.
> Sing, lads, and let us haste along.

In one of the more moving accounts of how songs punctuated everyday work, in 1862 a former slave who had lost 21 of her 22 children described how the work song "Poor Rosy," also known as "Heav'n shall-a be my home," should be sung.

In claiming that "Poor Rosy" was her favorite song, she also said, "it cant be sung widout a full heart and a troubled speerrit!" This work song could be heard through the act of rowing and other types of labor. The plaintive lyrics captured the troubled spirit alluded to earlier:

Poor Rosy
Poor Rosy, poor gal
....
Rosy, break my poor heart,
Heav'n shall-a be my home
I cannot stay in hell one day,
Heav'n shall-a be my home
I'll sing and pray my soul away
Heav'n shall-a be my home

Got hard trial in my way,
Heav'n shall-a be my home
O when I talk, I talk wid God,
Heav'n shall-a be my home

I dunno what de people want of me
Heav'n shall-a be my home[35]

More common, however, than providing employment was the river's role in carrying enslaved people to slave markets in New Orleans and Natchez. In numerous accounts ranging from self-emancipated slaves to the children of former slaves, the brutality of the trip downstream was articulated. Certain images prevailed, such as the one portrayed by Wm. J. Anderson in his *Life and Narrative.* In Anderson's life story, he was traveling to Tennessee where he was kidnapped and eventually taken to the Natchez slave pen. In his account, he recalled stopping at Nashville where more African Americans were picked up for the trip downstream to Natchez. In moving prose, Anderson captured the scene when he wrote, "At the raising of the river we were all huddled on the boat, very uncomfortably together, and down the river we went, still grieving, mourning and sorrowing over our fate." But their sorrow did not end there as an accident occurred, where in Anderson's words, "a gang of colored men, chained together were drowned in the hull of the boat." He went on to describe their cries and anguish as they were unable to break free. In describing his own despair, Anderson recalled at one point, "I sat down by the Mississippi River and wept." Finally, Anderson offered one of the best examples of how narratives about the river were influenced by race and class, producing a different riverine history for African Americans. In his reflections, he noted that the steamships were often referred to as "floating palaces" but to Anderson they represented a "dungeon."[36]

At the end of the trip downriver, African Americans faced another harrowing experience as they were auctioned off at slave markets bordering the river in cities such as Natchez and New Orleans. In New Orleans alone, 52 sites where enslaved people were sold existed; all these were located near the river. Numerous oral histories and slave narratives include passages about what William Brown called the "negro-pen" and the logistics in preparing enslaved people for auction. In another account by Issac Stier, he described his father's trip to the New Orleans "slave pen" where he compared their treatment with that of racehorses. Henry Watson, who successfully escaped slavery by way of a steamer down the Mississippi, recalled his experience at an auction block in Natchez. In Watson's retelling, he stressed how the "keeper" would

> grease the mouths of the slaves, so as to make it appear they are well and hearty, and have just done eating fat meat; though they seldom, if ever, while in the custody of the keeper, taste a morsel of meat of any kind.

Testimonies abound regarding the lengths that enslavers went to in securing the best prices for those enslaved, as former slave narratives provided detailed accounts of the auction process in one of the basest aspects of slavery. Again, the river served as the conduit for these markets where in some instances those enslaved did not even leave the ships as traders came on board.[37]

Echoing the experiences of Anderson and others, however, were those of others recounting the trip downstream. In their recollections, African Americans emphasized different aspects of the trip "down the river." For many, the most disturbing part was the breakup of families. In remembrances collected by the Federal Writers' Project from 1936 to 1938, Celia Henderson, born in 1849 into slavery in Kentucky, told of her mother being sold to pay off the master's debt. Leaving from Louisville to Natchez, Henderson spoke of the family being separated when her mother

> tuk us four children 'long wid her an pappy an the others staid
> back in Louisville. Dey tuk us all on a boat d de? Big Ribber—
> evah heah ob de big ribber? Mississippi its name—but we calls it de big
> ribber. . . .Natchez on de hill—dats whaah de tuk us. . . . No'em, nevah
> see pappy no moah....Im 'member mammy cryin' goin' down on de
> boat, and us chillum a cryin' too, but de place we got was a nice place.[38]

From the perspective of one left behind, William Brown related his feelings of despair when his mother was taken downstream to New Orleans. He recalled, "the boat moved gently from the wharf, and while she glided down the river, I realized that my mother was indeed "gone,—gone,—sold and gone, to the rice swamp dank and lone!" One of the most emotional accounts of family separation was written by Solomon Northup, a freeman who was kidnapped in 1841, impressed into slavery, and finally returned to his family and freedom in the North 12 years later. Abducted from his family, Northup was taken to New Orleans where he was sold to a planter with holdings on the Red River. In his memoir, Northup recalled the heartbreaking story of Eliza, who was separated from her children in New Orleans. She never saw them again but in Northup's words, they were never far from Eliza's thoughts. He lamented,

> In the cotton field, in the cabin, always and everywhere, she was talking of them—often to them, as if they were actually present. Only when absorbed in that illusion, or asleep, did she ever have a moment's comfort afterwards.

Unfortunately, memories such as these were not unusual as slave narratives and post–Civil War interviews exposed.[39]

Yet African Americans recognized the ambiguous roles that the river played: At one juncture, the river was the vehicle into the slave-pens in New Orleans or Natchez, whereas at another juncture the river was the means to escape. When the self-emancipated slave, John P. Parker, described his plans to escape North by way of a steamship from New Orleans, he elaborated, "There was a fascination about the river that I could not resist, because I knew that was my only avenue of escape from my bondage." and later wrote, "The Mississippi River attracted me like a magnet." But he also knew that the river could be his demise as determined if he was not successful, he would "go to the river for my final escape." For some, death was preferable to being sold down the river, as one woman went down to the river and drowned herself. For many determined to escape, New Orleans was often the starting point as John Sella Martin expressed: "New Orleans was the place I desired to reach, that I might conceal myself on some steamboat and come up the Mississippi River to the Ohio river, and so reach Ohio." Solomon Northup voiced the same sentiments as he attempted escape.[40]

Along with offering escape, the river's aesthetics were also appreciated. As Henry Bibb described the beauty of the Ohio River—in one passage equating the "lovely stream" with freedom—Solomon Northup recalled the landscape bordering the Red River. His poetic observations of the nature surrounding him evoked a mysterious but rich environment. Undoubtedly influenced by his home environment in New England, Northup wrote the following about one of the areas where he worked: "From every tree, moreover, hang long, large masses of moss, presenting to the eye unaccustomed to them, a striking and singular appearance." Later, he described the "fertile lowlands bordering the Red River." Another appreciation of the riverine landscape, echoing the ambiguity found in Bibb's *Narrative*, was found in Andrew Jackson's narrative. As Jackson recalled his journey from slavery to the Ohio River, he reflected upon a moment along the Wabash River where he "heard the merry birds singing in the branches over my head, and saw the bounding squirrels as they leaped from tree to tree," acknowledging their "ample domain," for "wild sports and songs." Like Bibb, Jackson lamented, "Would to God the ties of nature, were among men as they are among thee." Coinciding with references to the river's natural beauty, the arrival of steam invited another way of knowing the river where trauma and violence did not dominate. Memories included working on the longest steamship or the fastest along with one unique account regarding individual steamship whistles. In this story, written by Lafcadio Hearn in 1869, he recalled meeting the legendary Albert Jones, known for imitating the whistles of every steamboat on the Mississippi and Ohio Rivers. As Hearn recalled, Jones claimed he could discern the different whistles of incoming steamboats. In a demonstration, Jones

> suddenly threw up both hands, concave-fashion, to his mouth, expanded his deep chest, and poured out a long, profound, sonorous cry that vibrated through the room like the music of a steam-whistle. He started off with a deep nasal tone, but gradually modulated its depth and volume to an imitation of the steam-whistle, so astonishingly perfect that at its close every listener uttered an involuntary exclamation of surprise.

Hearn continued in a description of how Jones could even further refine the nuances in each steam-whistle. The aesthetics of the riverine environment and fond memories associated with steamboats, however, were spoken of less often than other associations. Instead, more common was the perspective of Josiah Henson:

> I know not what most men see in voyaging down the Mississippi. If gay and hopeful, probably much of beauty and interest. If eager merchants, probably a golden river, freighted with the wealth of nations. I saw nothing but portents of woe and despair. Wretched slave-pens; a smell of stagnant waters; half-putrid carcasses of horses or oxen floating along, covered with turkey buzzards and swarms of green flies,—these are the images with which memory crowds my mind. My faith in God utterly gave way . . . I saw only the foul miasmas, the emaciated frames of my negro companions.[41]

Not surprisingly, the memoirs of African Americans formerly enslaved were filled more with strategies of survival and searing observations of the cruelty of slavery.

By 1865, the institution of slavery had ended. With the abolition of slavery, Americans also celebrated a free Mississippi River. In 1863 after the Emancipation Proclamation, Abraham Lincoln stated, "The Father of Waters again goes unvexed to the sea." Yet, before freedom was experienced by all, memories of violence persisted. Anna Smith, formerly enslaved, recalled the Civil War years with the following scene:

> . . . the baying of blood hounds at night along the Ohio River, trying to follow the scent of escaping negroes and the crack of firearms as white people, employed by the plantation owners attempted to halt the negroes in their efforts to cross the Ohio river into Ohio or join the Federal army.

Up until the end of slavery, the memory of the riverine environment encompassed both freedom and enslavement. Further, the violent images that Smith projected persisted into the late nineteenth and early twentieth centuries despite the end of slavery. Now the wilderness, hinterland would be associated with baying hounds and lynchings—another collective memory of the Mississippi River and its environs. But the Civil War years also ushered in new associations as riverfront cities, such as Vicksburg, became beacons for freedom.[42]

For African Americans, post–Civil War life along the Mississippi River—whether laboring in levee camps, sharecropping on former plantations, or performing as roustabouts—remained an ambiguous experience. Technically free, working conditions were often little better than those when enslaved. The contradictory positions held by African Americans and whites persisted into the twentieth century, as the South mythologized a past that included serene plantation homes lining the river, supported by contented enslaved people. For both populations, past and present memories of the river at times converged, but they more often collided. Ironically, for the river, the same impulses that shackled African American labor also harnessed the river. By the early twentieth century, engineers envisioned a river free from its unpredictability—shifting channels, treacherous snags, and uneven depths. An African American labor force shaped the new river—building and maintaining levees, consigned to farm flood-prone bottomlands, and singled out for the most arduous work on the remaining steamships. Their experiences throughout the post–Civil War era built upon past associations while creating new memories of the "Father of Waters." The river represented not only a spiritual haven while a harsh taskmaster, a site for resistance but also a site for violence. These contradictions prevailed throughout the remaining nineteenth century and into the twentieth, crafting a Mississippi River history distinct from the narratives of enslavers, engineers, and politicians. For African Americans, the Mississippi River history conceived through myriad experiences—including work, escape, religion, and violence—is rich in the retelling as song, folklore, and prose chronicle the relationship.

Notes

1 Henry Bibb, *Narrative of the Life and Adventures of Henry Bibb, an American Slave, Written by Himself* (New York: MacDonald and Lee, Printers,1849; rpt. Echo Library, 2005):19.
2 Bibb, *Narrative*, p. 26.
3 Bibb, *Narrative*, p. 66.
4 In Wm. Brown's narrative, he commented on how the institution of slavery referred to those involved in the sale of enslaved people as "negro speculators," but African Americans called them "soul drivers." See Brown, *The Narrative of William W. Brown: A Fugitive Slave* (1847; rpt. ReadaClassic, 2010):25. Bibb, *Narrative*, pp. 52, 58.
5 Bibb, *Narrative*, p. 75.
6 For an expanded history of the development of the Deep South during the antebellum period, the significance of the Louisiana Purchase and Manifest Destiny, see Adam Rothman, *Slave Country: American Expansion and the Origins of the Deep South* (Cambridge: Harvard University Press, 2005); Mark Joy, *American Expansionism, 1763–1860: A Manifest Destiny* (London: Taylor and Francis, 2014); Peter J. Kastor and Francois Weil, eds., *Empires of the Imagination: Transatlantic Histories of the Louisian*

Purchase (Charlottesville: University of Virginia Press, 2009). The Louisiana Purchase resulted in 15 new states, doubling the size of the United States with 800,000 additional square miles. For further discussion focused upon the Lower Mississippi Valley, see Christopher Morris, *The Big Muddy: An Environmental History of the Mississippi and Its Peoples from Hernando de Soto to Hurricane Katrina* (Cambridge: Oxford University Press, 2012). An expanded cultural history of the Mississippi River during the antebellum era can be found in Thomas Ruys Smith, *River of Dreams: Imagining the Mississippi before Mark Twain* (Baton Rouge: Louisiana State University Press, 2007). Walter Johnson, *River of Dark Dreams: Slavery and Empire in the Cotton Kingdom* (Cambridge: Harvard University Press, 2012):40.

7 Johnson, *River of Dark Dreams*, pp. 5–7. For further discussion regarding the Deep South economy and its reliance on the institution of slavery, which, in turn, prompted a lucrative internal slave trade see Edward E. Baptist, *The Half Has Never Been Told: Slavery and the Making of American Capitalism* (New York: Basic Books, 2014).

8 John B. Cade, "Out of the Mouths of Ex-Slaves" *Journal of Negro History*, Vol. 20, No. 3 (July 1935):311; Benjamin Drew, *A North-Side View of Slavery: The Refugee: or, The Narratives of Fugitive Slaves in Canada, Related by Themselves, with an Account of the History and Condition of the Colored Population of Upper Canada* (Boston: J. P. Jewett & Co., 1856; rpt. General Books LLC, 2012):178. In one former slave account, he wrote, "I saw some of the dreadfulest treatment on the sugar farms in the sugar making season." See William A. Hall interview in Drew, *North-Side View of Slavery*, p. 84. Further testimony to the brutality of the slave regime in the Lower Mississippi Valley was supported by speculation that the mortality rate among slaves was so high that the planters would not allow census takers to count the slaves on their plantation. See John Hope Franklin and Loren Schweninger, *Runaway Slaves: Rebels on the Plantation* (New York: Oxford University Press, 1999):10. For a comprehensive account of life on the sugar plantations in the Deep South and all the jobs that enslaved people worked, see Richard Follett, *The Sugar Masters: Planters and Slaves in Louisiana's Cane World* (Baton Rouge: Louisiana State University Press, 2005). Charles Sealsfield, *The Americans as They Are: Described in a Tour through the Valley of the Mississippi* (London: Bradbury and Dent, 1828):57–58. Accessed at http://www.gutenberg.org/files/44268/44268-h/44268-h.htm

9 For an excellent discussion considering the river as a site of violence and trauma, see Cynthia Gooch, ""I've Known Rivers:" Representations of the Mississippi River in African American Literature and Culture," Ph.D. Dissertation, (University of Kentucky, 2019) accessed at https://uknowledge.uky.edu/cgi/viewcontent.cgi?article=1125&context=english_etds

10 One of the best overviews of the Deslondes Revolt can be found in Daniel Rasmussen, *American Uprising: The Untold Story of America's Largest Slave Revolt* (New York: HarperCollins, 2011). An overview of contemporary reactions to the revolt can be found in Thomas Marshall Thompson, "National Newspapers and Legislative Reactions to Louisiana's Deslondes Slave Revolt of 1811" *The Journal of the Louisiana Historical Association*, Vol. 33, No. 1 (Winter 1992):16. Although the Deslondes Revolt was one of the largest slave rebellions in the United States, it is not as well known as others prompting Gwendolyn Midlo Hall in an interview to reference this as "historical amnesia." See Marissa Fessenden, "How a Nearly Successful Slave Revolt was Intentionally Lost to History" *SmithsonianMag.com*, 8 January 2016, https://www.smithsonianmag.com/smart-news/its-anniversary-1811-louisiana-slave-revolt-180957760/. For an overview of slave revolts in Mississippi during the antebellum era, see Davidson Burns McKibben, "Negro Slave Insurrections in Mississippi, 1800–1865" *The Journal of Negro History* Vol. 34, No. 1 (January 1949):73–90.

11 Joe William Trotter, Jr., *River Jordan: African American Urban Life in the Ohio Valley* (Lexington: University Press of Kentucky, 1998):11. For an early history of commerce on the Mississippi River, see E.W. Gould, *Fifty Years on the Mississippi or Gould's*

History of River Navigation (St. Louis: Nixon-Jones Printing Co, 1889). Frank Donovan, *River Boats of America: From Flatboats to Floating Palaces, From the Cotton Blossom to the Staten Island Ferry* (New York: Thomas Y. Crowell Co., 1966):85. An overview of the 1854 Grand Excursion tour is provided in Jason T. Busch, et al., *Currents of Change: Art and Life Along the Mississippi River, 1850–1861* (Minneapolis: University of Minnesota Press, 2004).

12 Jacob N. Taylor, *Sketch Book of Saint Louis: Containing a Series of Sketches of the Early Settlement, Public Buildings, Hotels, Railroads, etc.*, Missouri State Historical Society; Thomas Bangs Thorpe, "Remembrances of the Mississippi" *Harpers New Monthly Magazine*, Vol. 12 (December1855–May 1856):35.

13 Frances Bremer, *The Homes of the New World—Impressions of America* (New York: Harper Brothers, 1853):174.

14 William E. Davis and F. Mark Kiernan, eds., "Biographical Sketch of the Life and Ancestry of George Forman of Stratford-Ontario-Canada," in "The Mighty Mississippi: Two 19th Century Accounts," *Louisiana History: The Journal of the Louisiana Historical Association*, Vol. 18, No. 3 (1977):347.

15 James Kennerly Diary (1827–1838), Missouri Historical Society Archives, St. Louis, Missouri; Wm. Carr Lane Papers, Anne E. Lane correspondence (13 March 1846), Missouri Historical Society Archives; *Sketch Book of Saint Louis*, Missouri Historical Society; *Louisiana Gazette*, 17 May 1805; *Louisiana Courier*, 18 July 1829; John Hope Franklin and Loren Schweninger, *Runaway Slaves: Rebels on the Plantation* (New York: Oxford University Press, 1999):27, 57, 112, 158; Stephanie LeMenager, "Floating Capital: The Trouble with Whiteness on Twain's Mississippi" *ELH*, Vol. 71, No. 2 (Summer 2004):405–431. For more discussion regarding the communication networks established through the river trade, see Thomas C. Buchanan, "Levees of Hope: African American Steamboat Workers, Cities, and Slave Escapes on the Antebellum Mississippi" *Journal of Urban History*, Vol. 30, No. 3 (March 2004):360–377. Although focused on maritime trade in the Atlantic, W. Jeffrey Bolster lent further evidence for the communication networks that evolved between seafarers and African American communities living in cities or plantations. See Bolster, *Black Jacks: African American Seamen in the Age of Sail* (Cambridge: Harvard University Press, 1974).

16 Thomas C. Buchanan, "Rascals on the Antebellum Mississippi: African American Steamboat Workers and the St. Louis Hanging of 1841" *Journal of Social History*, Vol. 34, No. 4 (July 2001):797–816; *Missouri Republican* (St. Louis), 10 July 1841.

17 Gould, *Fifty Years on the Mississippi*, pp. 228–229; Woodland Plantation Overseer's Diary, MSS 754, 15, 17–19, 28, 30 January 1861, 2 February 1861, New Orleans Historic Collection; George W. Johnson Woodland Plantation Correspondence, Folder 1, George W. Johnson to Bradish Johnson, 13 March 1838, New Orleans Historic Collection. Levee work dominated plantation life. On the Woodland Plantation, for example, which was one of the largest producers of sugar, molasses, and rum in the Delta, the overseers' diary had countless references to levee work, with entries such as "22 men at river on back levee" (19 January 1861) or "20 women with hand barrows making back levee" (30 January 1861). See "Woodland Plantation Overseer's Diary," MSS 754, 1861, The Historic New Orleans Collection—Williams Research Center.

18 The use of slave narratives became widespread by the 1960s, as scholars recognized their contributions in understanding the world enslaved people created. The autobiographies were written at great risk, jeopardizing the freedom of the self-emancipated slave, a further testimony to their veracity. See David Blight, "The Slave Narratives: A Genre and a Source" Gilder Lehrman Institute of American History accessed at http://ap.gilderlehrman.org/history-by-era/literature-and-language-arts/essays/slave-narratives-genre-and-source; John W. Blassingame, "Using the Testimony of Ex-Slaves: Approaches and Problems" *Journal of Southern History* Vol. 41, No. 4 (November 1975):473–492.

19 Anthony E. Kaye, *Joining Places: Slave Neighborhoods in the Old South* (Chapel Hill: University of North Carolina Press, 2007):109–110. In a review of Kaye's work, while Kathleen Hilliard criticized some of his conclusions, she also acknowledged his thesis regarding the place of neighborhoods that lent an "indirect brand of agency" which "recalibrated the balance of power between master and slave." See Hilliard, *Winterthur Portfolio* Vol. 43, No. 2/3 (Summer/Autumn 2009):262–263; Frederick Law Olmsted, *The Slave States* (New York: Capricorn Books, 1856; 1959):117; Roderick A. McDonald, *The Economy and Material Culture of Slaves: Goods and Chattels on the Sugar Plantations of Jamaica and Louisiana* (Baton Rouge: Louisiana State University Press, 1993):1, 69.

20 McDonald, *The Economy and Material Culture of Slaves*, p. 66; Terry Alford, *Prince Among Slaves: The True Story of an African Prince Sold into Slavery in the American South* (New York: Oxford University Press, 1977, rpt. 2007):60–61.

21 Timothy Flint, *A Condensed Geography and History of the Western States, or the Mississippi Valley*, Vol. 1 (Cincinnati: W.M. and O. Farnsworth, Jun. Printers, 1828):46, 62–63; John P. Parker, *His Promised Land: The Autobiography of John P. Parker, Former Slave and Conductor on the Underground Railroad* (New York: W.W. Norton, 1998):46–47. For further discussion regarding the changing attitudes toward swamps in the United States, see David C. Miller, *Dark Eden: The Swamp in Nineteenth-Century American Culture* (Cambridge: Cambridge University Press, 1989).

22 Solomon Northup, *Twelve Years a Slave* (1853; Auburn: Derby and Miller, rpt. 2014):95, 97; John W. Blassingame, ed., *Slave Testimony: Two Centuries of Letters, Speeches, Interviews and Autobiographies* (Baton Rouge: Louisiana State University Press):394.

23 Scholarship considering the role of alternate landscapes in the world of enslaved people is a growing field. A few notable works include Rhys Issac, *The Transformation of Virginia, 1740–1790* (Chapel Hill: University of North Carolina Press, 1999); J.M. Vlach, *Back of the Big House: The Architecture of Plantation Slavery* (Chapel Hill: University of North Carolina Press, 1993). In Vlach's work, he documented how landscapes were "reconceptualized" by those enslaved and argued, "Employing mainly behavioral strategies, slaves privately remapped the domains designed by planters, reconceptualizing their various assigned landscapes in ways that they found more suitable," p. 230. Adding to the literature is Stephanie Camp, in which she used Edward Said's term, "rival geographies." See Camp, *Closer to Freedom: Enslaved and Everyday Resistance in the Plantation South* (Chapel Hill: University of North Carolina Press, 2004). For a work that specifically examines different perspectives of swamp landscapes, see W.T. Cowan, *The Slave in the Swamp: Disrupting the Plantation Narrative* (New York: Routledge, 2005). Blassingame, *Slave Testimony*, pp. 394–395, 397; Peter Randolph, *Sketches of Slave Life from Slave Cabin to the Pulpit*, 2nd ed., ed. Katherine Clay Bassard (1855; Morgantown: West Virginia University Press, 2016):8. Randolph's text, in describing covert meetings in the swamps, observed:

> not being allowed to hold meetings on the plantations, the slaves assemble in the swamps, out of reach of the patrols. They have an understanding among themselves as to the time and place of getting together. This is often done by the first one arriving breaking boughs from the trees and bending them in the direction of the selected spot.

Randolph, p. 8. *Slave Songs of the United States* (New York: A. Simpson & Co., 1867):84. For further discussion of the meaning of wilderness for the enslaved, see Melvin Dixon, *Ride Out the Wilderness: Geography and Literature in Afro-American Literature* (Urbana: University of Illinois Press, 1987).

24 James McKaye, *The Mastership and Its Fruits: The Emancipated Slave, face to Face with His Old Master (Valley of the Lower Mississippi)* (New York: Loyal Publication Society, 1864):12, accessed at https://www.loc.gov/resource/lcrbmrp.t2313/?sp=12&r=-0.398,0.177,1.813,0.945,0

25 Kaye, *Joining Places*, p. 5; Issac, *Transformation of Virgina*, pp. 52–53; Vlach, *Back of the Big House*, p. 13. In Thadious M. Davis' work, *Southscapes*, the term "black spaces" is introduced but Davis sees these spaces, a product of racial exclusion, as also "creative arenas." See Davis, *Southscapes: Geographies of Race, Region & Literature* (Chapel Hill: University of North Carolina Press, 2011):17.

26 Shane White and Graham White, *The Sounds of Slavery: Discovering African American History through Songs, Sermons, and Speech* (Boston: Beacon Press, 2005):62; Jon F. Sensbach, "The Singing of the Mississippi: The River and Religions of the Black Atlantic" *Religion in North America: Gods of the Mississippi*, ed. Michael Pasquier (Bloomington: Indiana University Press, 2013):31. For an excellent discussion on the use of music to understand cultural differences, see Anne Ostendorf, *Sounds American: National Identity and the Music Cultures of the Lower Mississippi River Valley, 1800–1860* (Athens: University of Georgia Press, 2011).

27 During the Civil War years and immediately thereafter, there were several efforts to collect the spiritual and work songs of African Americans. One of the most well known was published by Thomas Wentworth Higginson, a Civil War Union officer and abolitionist. See Higginson, "Negro Spirituals" *Atlantic Monthly*, Vol. 116 (June 1867):685–694. Refer to John David Smith, "The Unveiling of Slave Folk Culture, 1865–1920" *Journal of Folklore Research*, Vol. 21, No. 1 (April 1984):47–62; John Lovell, Jr., The Social Implications of the Negro Spiritual" *Journal of Negro Education*, Vol. 8, No. 4 (October 1939):634–643 for overviews and insights of the post–Civil War efforts of Higginson and others.

28 *Slave Songs of the United States* (New York: A. Simpson & Co., 1867):84–85; Thomas C. Buchanan, *Black Life on the Mississippi: Slaves, Free Blacks, and the Western Steamboat World* (Chapel Hill: University of North Carolina, 2004):8; David McD. Simms, "The Negro Spiritual: Origins and Themes" *The Journal of Negro Education*, Vol. 35, No. 1 (Winter 1966):38.

29 Henry Edward Krehbiel, *Afro-American Folksongs: A Study in Racial and National Music* (New York: G. Schirmer, 1914):89; Melville J. Herskovits, *The Myth of the Negro Past* (Boston: Beacon Press, 1941, rpt. 1990). Herskovits' conclusions have not been without its critics; for a more recent and fuller treatment of Herskovits' thesis and its significance, see Jerry Gershenhorn, *Melville J. Herskovits and the Racial Politics of Knowledge* (Lincoln: University of Nebraska Press, 2004).

30 Albert J. Raboteau, *Slave Religion: The "Invisible Institution" in the Antebellum South* (New York: Oxford University Press, 2004):215; In discussing the place of religion in African American culture during times of slavery, the influence of African religious practices is undisputed. A derivative of these African origins is the significance of water. Although today many scholars have questioned Herskovitz's conclusions about the far-reaching legacy of rivers, his contributions remain a starting point for later scholarship. According to Walter F. Pitts, Jr., one of the attractions of the Baptist faith was "the rite of total immersion baptism . . . Total immersion closely resembled African water rites." Pitts, *Old Ship of Zion: The Afro-Baptist Ritual in the African Diaspora* (New York: Oxford University Press, 1993):45. Additional insightful works on the role of rivers include Bryan T. Sinclair, "Merging Streams: The Importance of the River in the Slaves' Religious World" *Journal of Religious Thought*, Vol. 53, No. 2 (1997):1–19.

31 William Tynes Cowan, *The Slave and the Swamp: Disrupting the Plantation Narrative* (New York: Routledge, 2005):125; Northup, *Twelve Years a Slave*, pp. 119–120; Parker, *Promised Land*, p. 30; Janie Scott Interview, Federal Writers' Project: Slave Narrative Project, Vol. 12, Alabama (14 July 1937):338 accessed at https://www.loc.gov/resource/mesn.010/?sp; Sally Murphy Interview, Federal Writers' Project: Slave Narrative Project, Vol. 1, Alabama (1937):291 accessed at https://www.loc.gov/resource/mesn.010/?sp=300&st=text; Vlach, *Back of the Big House*. For further discussion on the relationship between slaves and the surrounding environment, see

Rebecca Ginsburg, "Freedom and the Slave Landscape" *Landscape Journal*, Vol. 26 (2007):1–7.

32 *Maroon Societies: Rebel Slave Communities in the Americas*, ed. Richard Price (Baltimore: Johns Hopkins University Press, 3rd ed., 1996):161–162; James McKaye, *The Mastership and Its Fruits: The Emancipated Slave, face to Face with His Old Master (Valley of the Lower Mississippi)* (New York: Loyal Publication Society, 1864):12, accessed at https://www.loc.gov/resource/lcbmrp.t2313/?sp+12&r_0.398,0.177,1.813,0.945.0; Syliva Diouf, *Slavery's Exiles: The Story of the American Maroons* (New York: New York University Press, 2014):33, 131; Octave Johnson Interview in John W. Blassingame, ed., *Slave Testimony: Two Centuries of Letters, Speeches, Interviews, and Autobiographies* (Baton Rouge: Louisiana State University Press, 1977):394–395. Public mention of maroons and runaways was frequent during the antebellum period with entries in newspapers such as *The Liberator* in 1849, commenting upon flooding in New Orleans, which "forced maroons in nearby swamps" to enter the city at night to "pilfer." See Larry Gara, *The Liberty Line: The Legend of the Underground Railroad* (Lexington: University Press of Kentucky, 1991):29. For scholarship on maroons, see Herbert Aptheker, "Maroons within the Present Limits of the United States" *The Journal of Negro History*, Vol. 24, No. 2 (April 1939):167–184. Aptheker claims that from 1672 to 1864, 50 maroon communities existed "in various places and various times," p. 167.

33 Alisha J. Hines, "Geographies of Freedom: Black Women's Mobility and the Making of the Western River World, 1814–1865" Ph.D. Dissertation (Duke University, 2018):124; George Taylor Burns Interview, *Federal Writers Project, Slave Narrative Project*, Vol. 5, pp. 37–39 accessed at https://www.loc.gov/resource/mesn.050/?sp=43; *From Tennessee Slave to St. Louis Entrepreneur: The Autobiography of James Thomas*, Loren Schweninger, ed. (Columbia: University of Missouri Press, 1984):108. In T. Buchanan's work, *Black Life on the Mississippi: Slaves, Free Blacks, and the Western Steamboat World*. (Chapel Hill: University of North Carolina Press, 2004), he argued that work on Mississippi River steamboats allowed for more mobility and opportunities for escape for those enslaved. In Gooch's 2018 dissertation (cited above), she challenged his conclusions and concluded that "While it seemed that access to the river would offer the chance at escape—or at least improve the conditions of slavery—this was rarely the case." (p. 36) For support of Buchanan's work, however, consult J.H. Franklin and L. Schweninger, *Runaway Slaves: Rebels on the Plantation* (New York: Oxford University Press, 1999). In their review of runaway advertisements, they cited numerous instances in which slaveholders warned that runaway bondspersons would attempt to reach the river. Also supporting Buchanan's thesis is the dissertation by Alisha Hines, in which she looked at court records, newspapers, and other primary sources from the Middle Mississippi Valley region and found that the nineteenth-century evolving capitalist economy exposed black women to the legal possibilities of attaining freedom. In river cities, such as St. Louis, black women were exposed to contemporary ideas, people, and public spaces, which facilitated increased opportunities for economic mobility and in some cases, freedom. See Hines, "Geographies of Freedom."

34 George Merrick, *Old Times on the Mississippi*; Josiah Henson, *Father Henson's Story of His Own Life* (n.p., 1858; reprint 2016):51–52; Allen Sidney, "Allen Sidney Tells the Story of His Life" *Slave Testimony: Two Centuries of Letters, Speeches, Interviews, and Autobiographies*, ed. John W. Blassingame (Baton Rouge: Louisiana State University Press, 1977):525; James Hebron Moore, "Simon Gray, Riverman: A Slave Who Was Almost Free" *Mississippi Valley Historical Review*, Vol. 49, No. 3 (December 1962):472–484.

35 George Washington Cable, "Creole Slave Songs" *Century Magazine*, Vol. 31 (April 1886):807–828 in Bruce Jackson, ed., *The Negro and His Folklore in Nineteenth-Century Periodicals* (Austin: University of Texas Press):62, 239.

36 William J. Anderson, *Life and Narrative of William J. Anderson, Twenty-Four Years a Slave* (Chicago: Daily Tribune Book and Job Printing Office, 1857; rpt. 2016).

37 Gooch, "I've Known Rivers," p. 23; In the New Orleans slave market, the former slave William Wells Brown recalled that one of his jobs was working for a "soul driver" and preparing slaves for sale. One of the jobs he was tasked with included applying a black dye to their skin to make them appear younger. P. 21, Fergus M. Bordewich, *Bound for Canaan: The Epic Story of the Underground Railroad, America's First Civil Rights Movement* (New York: Amistad, 2005); Issac Stier, Federal Writers' Project: Slave Narrative Project, Vol. 9, Mississippi (1936–1938):143 accessed at https://www.loc.gov/resource/mesn.090/?sp=15; Northup, *Twelve Years a Slave*, p. 38; Henry Watson, *Narrative of Henry Watson: A Fugitive Slave* (Boston: Bela Marsh, 1850):16.

38 Celia Henderson Interview, Federal Writers' Project: Slave Narrative Project, Vol. 12, Ohio, (nd):42 accessed at https://www.loc.gov/resource/mesn.120/?sp=46.

39 William Wells Brown, *The Narrative of William W. Brown: A Fugitive Slave* (1847; ReadaClassic, 2010, rpt.):46; Northup, *Twelve Years a Slave*, p. 55.

40 John P. Parker, *His Promised Land*, ed. Stuart Seely Sprague (New York: W.W. Norton and Co., 1996):36, 39, 63; John Sella Martin Interview in Blassingame, *Slave Testimony*, p. 729; Northup, *Twelve Years a Slave*, p. 41.

41 Northup, *Twelve Years a Slave*; Andrew Jackson, Narrative and Writings of Andrew Jackson, of Kentucky (Syracuse: Daily and Weekly Star Office, 1847; rpt. 2015):16; Lafcadio Hearn, *Children of the Levee* (1876; Lexington: University of Kentucky Press, rpt. 1957):8–10; Josiah Henson, *Father Henson's Story of His Own Life* (n.p., 1858; reprint 2016):60. For additional insights on how African Americans viewed their environment see Dianne D. Glave, *Rooted in the Earth: Reclaiming the African American Environmental Heritage* (Chicago: Lawrence Hill Books, 2010).

42 Anna Smith Interview, Federal Writers' Project: Slave Narrative Project, Vol. 12, Ohio (11 June 1937):85 accessed at https://www.loc.gov/resource/mesn.120/?sp=89.

3

THE POST-WAR YEARS

New River Roles

FIGURE 3.1 Refugees on Levee, April 17, 1897.
Credit: Library of Congress Prints and Photographs Division.

On May 22, 1879, the *New York Times* reported hundreds of African Americans waiting on Mississippi River landings stretching from Greenville, Mississippi to a point 160 miles north of Vicksburg (Figure 3.1). They were waiting for steamers to stop and carry them to St. Louis where they would continue west in

DOI: 10.4324/9781315617077-4

search of Kansas farmland. Known as the Exodusters, these nineteenth-century emigrants looked to Kansas as an escape from the harsh conditions in the post–Reconstruction Southern states. Beginning with a few hundred emigrants in March 1879, by 1880 their numbers reached from 15,000 to 20,000.[1] While a diverse group, all shared a common goal: to pursue lives free from white Southerners' intent on recreating the antebellum South. For example, one African American man in his seventies told a *New York Times* correspondent:

> Forty years I was taken South in irons, not that I was a criminal or vicious, but that I was a slave. I have remained here during the war, through pestilence, and the horrors, and have never failed to assist in making a crop. I have borne the evils of bad masters, civil and military, but the present system of working and disposing of crops is such that I can't live here. No master ever exacted of me the labor necessary now to keep out of debt. Year after year we have fewer luxuries, until it is impossible to provide warm clothing in Winter and medicine in Summer. I know it is not because I don't earn it, for we make good crops of cotton and live as cheap as possible, but the books always show me in debt.

He went on to say that when complaints are lodged, landowners respond with "curses and threats, and sometimes blows." In contrast, he maintained, "Kansas has no terrors for us."[2]

Still another Exoduster, John Solomon Lewis, recalled bringing his family to Kansas and "free ground." Upon their arrival in 1879, Lewis suggested:

> a little prayer meeting; so we held a little meeting on the river bank. It was raining but the drops fell from heaven on a free family, and the meeting was just as good as sunshine. We were thankful to God for ourselves and we prayed for those who could not come.[3]

The idea of "free ground" motivated the Exodusters, even if after reaching Kansas their circumstances remained strained as most were without resources. Yet, the idea of returning to the Southern states was not an option as one memorial sent to Washington, D.C. reported: "Whatever becomes of them . . . They are unanimous in the unalterable determination not to return." Or in the words of one woman, waiting on the levees surrounding St. Louis for transportation to Kansas, "We'se goin' to Kansas and we won't go back dar."[4]

As word spread about supposed opportunities in Kansas, more African Americans from the Lower Mississippi Valley inquired as to conditions in the state. In one letter addressed to the governor in Topeka from African American citizens living in the Louisiana Parish of Iberville, the governor was informed that a mass meeting had been held for the purpose of sending a committee to Kansas next March. The committee was to "investigate into the condition of the state and report back to us whether it would be favorable for us to emigrate to that state or not." The author of the letter impressed upon the governor their credentials as a hardworking "laboring class" with many owning property and stock. Their reasons for leaving included not being adequately paid for their crops in Louisiana. The letter, one of several, revealed a deliberate, thoughtful movement by groups intent on a better life. Letters such as these, combined with reports of makeshift camps—located by St. Louis—for those awaiting passage to Kansas, revealed an ongoing resilience and self-determination on the part of African Americans living along the Mississippi River. The resilience shown by African Americans in the antebellum period, despite living under a regime of slavery—persisted through movements such as the Exodusters.[5]

As the personal letters indicated, one reason for the mass emigration to Kansas with its promise of a better life were intolerable working conditions in states, such as Mississippi and Louisiana, where the greatest number of Exodusters originated. The Compromise of 1877, resulting in the election of President Samuel R. Hayes, ended the Reconstruction Era and its reconfiguration of a Southern political landscape. The Reconstruction Era witnessed an active Republican Party in the South, with African American participation in state and national politics—resulting in 16 African Americans serving in Congress—along with passage of the Thirteenth, Fourteenth, and Fifteenth Amendments and the Civil Rights Acts of 1866 and 1875. These gains—although land distribution never occurred and in some places land ownership by African Americans was forbidden—would be eclipsed with the rise of the Southern Democrats throughout the South, negating Reconstruction gains. (Land ownership was often contested as towns such as Opelousas, Louisiana, which forbid African Americans "to rent or keep a house within the limits of the town *under any circumstances*, and any one thus offending shall be ejected and *compelled to find an employer* or leave the town within 24 hours.)" Still, the period once signaled hope, as W.E.B. DuBois characterized Reconstruction as an effort "to make black men American citizens" yet ultimately "was in a certain sense all a failure, but a splendid failure." By the end of the nineteenth century, not only would Reconstruction's successes disappear but also the world of a Jim Crow South with all its travesties would be established. As a result, life for African Americans still residing in the South changed dramatically after 1877.[6]

Throughout Southern states, Republican office holders were voted out and replaced with a white-led Democratic party. In numerous instances, attempts by African Americans to either run for office or vote for a Republican candidate were met with violence through a practice labeled "bulldozing," silencing African Americans in these states. Bulldozers were white terrorists known for their intimidation tactics that ranged from warnings or threats to lynchings. Thus, African American voting rights were constantly undermined as these terrorists worked in tandem with other white Southerners for a return to a pre–Civil War political and social environment. Alarmed by the viciousness of the bulldozer attacks, often resulting in African American deaths, many African Americans feared a return to slavery. Their alarm was not far-fetched. For example, in 1875,

General H. Sheridan, who was in command at New Orleans, reported: "There is ample evidence, however, to show that more than twelve hundred persons have been killed and wounded during this time on account of their political sentiments." He went on to cite several parishes as the sites of "frightful massacres."[7]

Supporting Sheridan's remarks, and in the same 1880 *Senate Report*, were numerous firsthand accounts chronicling the violence occurring in many of the Southern states. Henry Adams, a Civil War veteran, who after the war devoted himself to documenting the "crimes against his race," in the states of Louisiana, Mississippi, and Texas, reported the names of 683 "colored men who have been whipped, maimed, or murdered within the last eight years." The *Senate Report* uncovering these abuses was responding to the mass exodus of African Americans from North Carolina, in addition to the Mississippi Valley states. Seeing large numbers of African Americans leaving the South, white Southerners became increasingly alarmed that they might lose a labor force. In reaction to their concern, Congress held hearings to determine why so many African Africans were leaving their homes for Kansas. By late 1879, the Senate appointed a committee to investigate the growing number of African Americans, commonly known as Exodusters, bound for Kansas. The report included a minority statement concluding that African Americans leaving the Southern states, particularly Louisiana and Mississippi, which included the greatest number of Exodusters, were prompted by a tenancy system that amounted to debt peonage and numerous acts of violence. After listening to the testimonies of hundreds, the minority report concluded that the Exodusters left the Southern states due to "great privation and want from excessive rent exacted for land, connected with murder of colored neighbors and threats of personal violence to themselves." As mentioned earlier, the violence, or bulldozing, resulted when African Americans sought to run for office or vote on the Republican ticket.[8]

The story of the Exodusters—following a historical trajectory of self-determination and resilience in African American history—captured the predominant themes for African Americans living along the Mississippi River as the river served as a freedom highway north. Since the majority of Exodusters came from Mississippi and Louisiana, they relied upon the Mississippi River and its majority tributary, the Missouri River, for entry into Kansas. As in the days of slavery, the river represented a means to escape as observers noted hundreds waiting on its riverbanks for steamers headed north to St. Louis. Once boarded, the journey upriver to Kansas could take up to eight days. Even the trip upstream the Mississippi River bore resemblances to earlier antebellum journeys downstream. Although not in chains, the traveling conditions—sleeping and eating on deck with no amenities—for African Americans harkened back to antebellum days. In this historical moment, the river again integrated conflicting narratives as it was both a means to freedom and often a dehumanizing journey.[9]

Upon reaching St. Louis, the river and its environs remained a constant in the next chapter of African American history with its familiar themes of displacement and resilience. The levees—always part of African Americans' experience with the river—assumed new imageries as sites not only associated with labor but also associated with refuge. Throughout the brief history of the Exodusters, reports, often accompanied by photographs, circulated of African Americans, surrounded by their belongings, waiting on Mississippi River levees for passage to Kansas. Beginning in the 1870s, a collective memory emerged where Mississippi River levees represented two imaginaries—as places of escape whether from the mistreatment of white Southerners or ravaging floods juxtaposed with associations of labor. By the late nineteenth century, working on the levees became associated with the notorious levee camps where a convict lease labor force, drawing primarily upon African Americans, worked under punishing, life-threatening conditions. For the Exodusters, however, the levees—a product of African American labor dating back to the founding of New Orleans—became sites for transitional homes as makeshift camps appeared on the levees protecting the city. In this instance, the levees had evolved into sites of refuge while the river facilitated escape.[10]

The Exodusters remained on the levees until the last leg of their journey—securing transport on the Missouri River to Kansas and the Promised Land. Songs were sung commemorating the movement that assumed the trappings of a spiritual journey. Even the name, Exodusters, referenced the Biblical Exodus and the Mississippi River continued its historical role as a spiritual host to African Americans. As in antebellum times when hymns such as "Crossing Over Jordan" might reference the enslaved person's journey to freedom, the Exodusters relied upon the Mississippi River to free them from the tyranny of a resurgent white South. The song, "The Land That Gives Birth to Freedom," written in 1877, memorialized the Exoduster movement and one of its leaders, Pap Singleton. The lyrics captured the spiritual aspects of the migration as well as the conditions that prompted the move:

> We have held a meeting to ourselves, to see if few can't
> Plan some way to live.
>
> CHORUS
> Marching along, yes we are marching along,
> To Kansas City we are bound
>
> We have Mr. Singleton for our President, he will go on
> before us and lead us through.
>
> Surely this must be the Lord that has gone before him,
> and opened the way.
>
> For Tennessee is a hard slavery State, and we find no
> Friends in this country.
>
> Truly it is hard, but we all have to part, and flee into a
> Strange land unknown.

....

We are on our rapid march to Kansas, the land that gives
Birth to freedom. May God Almighty bless you all.
Farewell, dear friends, farewell.

....

The time has come, we all must part, and take the parting band,

It seems to me like the year of jubilee has come; surely
this is the time that is spoken of in history.
Farewell, dear friends, farewell.[11]

However, by 1880, despite the fanfare and white anxieties, the majority of Exodusters had completed their journey to Kansas where they settled in towns such as Wyandotte or the recently established African American town of Nicodemus. The state's African American population had increased dramatically over a two-year period as Kansas went from a population of 17,108 African Americans in 1870 to 43,107 African Americans in 1880. While Nicodemus did not experience the prosperity that was envisioned and many emigrants did not realize the promises of migration, the short-lived movement revealed a mobility and determination through the agency of the river and its surroundings dating back to the 1700s. Integral to the narrative of mobility and determination was the evolving imagery of the levees. As mentioned earlier, beginning in the early 1700s, the levees were sites of labor, protecting New Orleans and the growing number of riverfront plantations from flooding. But by the 1870s, the levees—through the actions of the Exodusters—assumed an iconic status as sites of refuge as well as labor, which persisted throughout the twentieth century. Through movements such as that of the Exodusters, the post-war years were formative ones in shaping twentieth-century memories of the Mississippi River and its surrounding levees. For example, the 1927 Flood would be another pivotal moment in the collective memory of African Americans in their long history with the river and its surrounding levees.[12]

For African Americans, however, who did not participate in the Exoduster movement labor in the South was not limited to sharecropping or tenancy. After the Civil War, the once-prosperous South was in shambles. The carefully maintained levees—corseting the river, ensuring a navigable river along with lessening the catastrophic effects of flooding—were in disrepair and would play another role in the lives of African Americans. To return to a pre-war economy, or for many white Southerners to develop a modernized South, Southern planters and politicians recognized the need for infrastructure and public works. Navigation and flood control were critical in transforming the post-war South, as Southern leaders sought to impose order on an unruly Mississippi River with the construction of levees being a primary focus. By the 1870s, regaining control over the Mississippi through an extensive levee system lining the riverbanks became a panacea for Southern economy recovery. As discussed in Chapter 2, up until the outbreak of the Civil War, individual planters were responsible for most levee construction and maintenance through the labor of the enslaved. Each state bordering the river also assumed some responsibility, but these efforts were piecemeal with no central oversight. After the Civil War, when the levees were in disrepair and flooding threatened crops, planters, an emerging urban class, and Southern politicians, seeing the advantages of centralization, asked the federal government to intervene and provide assistance.[13]

Supporting their claims was a report from the U.S. Army Corps of Engineers Chief Engineer A. A. Humphreys, in which he noted that in Louisiana alone, 108 miles of levees were in disrepair. White Southerners continued to ask Congress for internal improvement funding for the levees and railroads and eventually their requests were granted when in 1879, the Mississippi River Commission was formed. The Commission was charged with flood control and navigation, returning the river to its historical place as one of the nation's primary arteries. With the Commission in place, oversight of the construction and maintenance of levees was centralized and the system that bordered the river for the next several decades was built. The history of the river changed dramatically at this juncture as the Corps and its cadre of professionals foresaw an engineered Mississippi River that would be subdued and predictable, facilitating navigation and vanquishing floods. By the twentieth century, their assumptions would be challenged as the Great Flood of 1927 demonstrated. In the meantime, however, the fallacy of containment retained a strong hold on the engineering community and the white Southerners who enlisted their aid. In contrast to this worldview, for African Americans who lived and worked on the river, the challenges of an arbitrary, unpredictable Mississippi did not diminish. Through their labor, such as building levees, or farming the first-to-flood bottomlands, the river retained its agency. It would be no accident that the Great Flood of 1927 resulted in differing experiences for African Americans and white Southerners.[14]

But in returning to the work of building and maintaining the levees along with other major public works, white Southerners recognized the need for a permanent labor force. If the post-war South were to build a modern-day infrastructure, a dependable, guaranteed labor force was critical. Prior to the Civil War, those enslaved performed most of the labor. Intent not to lose this labor force in the development of a New South and preserve a white supremacist hierarchy, Southern states began passing laws circumscribing the lives of African Americans and ensuring their presence in the South. These restrictive laws resulted in slave-like conditions for many African Americans. As legal protection for African Americans diminished in states such as Mississippi and Louisiana, minor offences such as vagrancy, missing work for a certain number of days, and property theft were often punishable by incarceration. For example, in Mississippi, the "pig-law" was passed, which allowed the courts to incarcerate any offender who stole more than $10 worth of property. Incarceration could carry a five-year sentence. By imprisoning African Americans on these trumped-up charges, states such as

Mississippi and Louisiana had a ready-made labor force for their much-needed public works, particularly the levees and railroads. Although the related goals of constructing public works and securing a labor force represented a new departure for white Southerners—as they envisioned a modern state, replete with a robust economy and its supporting infrastructure—for African Americans, the post-bellum chapter was a continuation of a history with the river where the river represented violence and trauma juxtaposed with migration, spirituality, and escape. Although slavery ended in 1865, for African Americans the back-breaking work of building and maintaining levees persisted into the twentieth century. By 1865, African Americans had been constructing levees on the river's banks for more than 150 years. Ironically, their history through slavery and freedom was entwined with Southern efforts to subdue the river in tandem with subduing the workforce.[15]

However, building and maintaining levees was not the only river-related labor during the period from 1865 to the 1920s. With the end of slavery, many African Americans left for the rich, relatively untapped Yazoo Delta lands. For a brief period during the 1870s to 1880s, land ownership was realized in the area. But by the turn of the century, the predominant arrangements for African Americans farming the rich Delta lands were either tenancy or sharecropping. This partnership with white Southern landowners severely hampered African American land ownership, instead resulting in lifelong debt for the tenant or sharecropper. Still, another type of river-related labor—dating back to the antebellum period—was working on the steamboats. Up until the end of steamboat travel, African Americans labored on steamboats and the piers in numerous jobs. One of the more popular, however, was the job of a roustabout. As roustabouts, African Americans again performed some of the most difficult jobs. Never well paid, the roustabouts lived impoverished lives that would be memorialized in song. While other jobs connected to the river and its environs were recorded during the post-war years such as clearing swamplands, focus will remain on levee construction and maintenance, farming, and "rousting." Through these three jobs, joined with earlier associations with the river, African Americans' collective memory of the river expanded. New imageries of levees, memories of "rousting," and the debt-ridden system of sharecropping and tenancy added to earlier memories of the river's role in African American lives as their intertwined histories expanded. As these new imageries and memories of the river emerged, they were reflected in the cultural outlets of song, folklore, and narrative. Underlying the new imageries of the river, however, the same themes prevailed resilience, displacement, and mobility as the river offered refuge, solace, and labor—while alternating between sites associated with violence and freedom.[16]

Returning to the task of building and maintaining levees, once the Corps of Engineers received approval and funding to construct a levee system along the length of the river, they subcontracted much of the labor. The states, in turn after the Civil War, utilized a growing prison population from the state penitentiary to perform the back-breaking work associated with the levees. Although levee camps were also populated by African Americans, not incarcerated, as they looked for additional income. But this system of using prisoners for public works, such as levees and railroads, known as the convict lease system was prevalent in the states bordering the Mississippi River. Most of the prisoners were African American. For example, by 1893, out of a total 1,090 prisoners in Louisiana, African American males made up 89% of the population. Work conditions were brutal, prompting officials to insist that only the most physically fit were to work in the levee camps—a provision often ignored. According to official accounts, no prisoner survived a ten-year sentence, prompting several critiques of the convict lease program. One regional newspaper reported, "The average convict life is six years. It would, therefore, be more humane to punish with death all prisoners sentenced to a longer period than six years." Although reconfigured in states such as Louisiana by the end of the nineteenth century, the system still lasted until 1933.[17]

In Louisiana, for example, the convict lease system was flourishing after the Civil War and did not begin to decline until the last decade of the nineteenth century. For the lessee, the arrangement was very profitable as the most well-known lessee, Colonel Samuel Lawrence James' profits reveal. In 1869, James amassed almost $100,000 from work performed by convicts on the levees. In 1870, he realized similar profits. The profits, however, came at the expense of African American lives. In an 1884 critique of the convict lease system, the reporter noted: "This remorseless cruelty of the present management has been more fatal to the lives of the convicts than the percentage of losses of the two armies during the civil war." For example, in 1881, the mortality rate for convicts was 14%, despite the requirement that only those most physically fit should work on the levees. Other newspaper articles followed, exposing the system and its brutality. Some of the articles were particularly graphic in describing the physical condition of the prisoners or their surroundings with accounts of needless amputations, scarred bodies, and vermin-infested hospitals. Although the state eventually took over the convict lease system, contracting labor out to public works, James leased convicts for almost three decades. Under James' brutal regime, the workforce—comprised primarily of male African Americans—labored on the levees, railroads, and later his cotton plantation, Angola. In Mississippi, Angola's counterpart would be the notorious Parchman Farm.[18]

While the convict lease system represented new institutional arrangements, for African Americans, the levee camps under the convict lease system were an extension of earlier associations with the river. But there were differences, resulting in worse physical conditions for African American labor. Since the lessees, such as James, did not "own" the labor force, one Southerner observed, "But these convicts, we don't own 'em. One dies, get another." Under the convict lease system—with its new levels of brutality where armed guards supervised the work crews and some critics contend actually encouraged violence among the workers—new sites of violence were introduced along the river, as the high mortality rates at the levee camps disclosed. According to one late nineteenth-century critic, the mortality rates under the convict lease system were greater than those during the plague. The same critic claimed that the lease system was the worst in Louisiana, Mississippi, and Arkansas—all states bordering the Mississippi River. Thus, the punishing work of levee building in the late nineteenth century—which continued into the twentieth century—reached new heights as the levees became symbols of modernization with the South's drive to restrain the river and master the environment. For many, the irony was not lost, harnessing a river that paralleled the continued subjugation of African Americans. By the twentieth century, years of levee work would be commemorated in song as blues artists lamented the back-breaking work in the camps.[19]

In addition to public critics of the convict lease system, African Americans also documented levee camp conditions where violence insured productivity. These firsthand accounts were drawn from African Americans who were not incarcerated but looking for employment. Whether incarcerated or "free labor," working conditions at the levee camp would be familiar as the jobs of muleskinners or camp tenders, for 14 hours a day, were back-breaking, torturous work. Again, by the late nineteenth century, most levee workers were African Americans as it was not unusual for the camp to comprise three armed white supervisors directing the work of 100 black men and women. Accompanying the violence were the inhumane work conditions, such as the following firsthand account by Big Joe Williams. Although written in the twentieth century, his recollections still held true for earlier levee workers. In Williams' account, he recalled working on the levees at the age of 12:

> I left home run off to the levee camp. I was about twelve years old then. I went to a camp in Greenville, Mississippi . . . I went out there and was a willow driver. Yeah, popped lossa mules out there . . . mule driver. The life was hard. The men worked from sunrise to sunset. At night they slept in filthy tents on rotten mattresses with a couple of blankets to crawl under. The food just about kept a man alive . . . The pay was $1 to $1.50 a day and that went on Saturday-night drinking and women.[20]

In another particularly graphic firsthand account, a levee worker who worked as a muleskinner recalled,

> And at night you better clean the collars of your mule team. If you didn't clean the collars off, next morning old boss go out there and see a scab on that mule's collar and he'd make you eat it off.

For those levee workers working outside the penal system, another injustice was a credit system in which the worker could never pay off his debt for supplies.[21]

Lasting up until the 1930s when levee construction became more mechanized, the levee camps produced their own cultural norms. Part of the cultural milieu comprised the work songs associated with the levees. Later known as the

Delta blues, these songs, many of which were written in the twentieth century, provided insights into camp life, reinforcing the firsthand accounts of the camp brutality. For example, in a song by Washboard Sam, titled "Levee Camp Moan," he illustrates a levee worker's life. He bemoans a diet consisting of beans in the morning and night, working all day and night, and hands that were sore as he "wired so many wagons." Adding to the abysmal conditions was the uncertainty of a paycheck or in the singer's words, "I never did know just when we were due our pay."[22] Women also lamented life in the levee camps. In a well-known song by Lucille Bogan, written in 1927, she offers a glimpse into levee life for a woman:

"Levee Blues"'

Down on the levee, Camp Number Nine
Down the levee, Camp Number Nine
You can pass my house, honey you can hear me cry
I never had no blues, until I come by here
I never had no blues, until I come by here
I'm going to leave this camp, you can't start in here
My sister got them, brother got them too
We all got the levee camp blues
I ain't found no doctor, ain't no doctor in this whole round world
I ain't found no doctor, ain't no doctor in this whole round world
Just to cure the blues, the blues of a levee camp girl.[23]

Songs such as these, which became popular in the twentieth century, were one way to retrieve a collective memory regarding the river and how the river was perceived by African Americans. The songs illuminate a number of themes associated with the levee camps: working conditions, separation from family, violence, and disorder.

Thus, Mississippi River levees became a constant in the collective memory of African Americans as the Corps of Engineers remained committed to an engineering strategy whereby levees continued to be built on the belief that the Mississippi River could be contained. However, nineteenth-, twentieth-, and twenty-first century floods, such as the Great Flood of 1927 and the aftereffects of Hurricane Katrina, exposed this fallacy but not before many African Americans lost their lives and/or their livelihoods. In the 1927 Flood, the unofficial death toll reached 1,000. Of those 1,000, at least 100 African American men drowned on the night that National Guard troops forced them to remain at the Mounds Bayou levee in a last-ditch effort to save the levee. For Hurricane Katrina, the death toll reached almost 2,000 with disproportionate losses in African American communities.[24]

Another labor-related link to the river was derived from the work of the roustabouts loading and unloading the steamers. Before the Civil War, roustabout work was undertaken primarily by immigrant groups, such as the Irish and

Germans. With the end of slavery, however, as immigration declined and African Americans provided a cheaper source of labor, the roustabout community became associated with African Americans. This new age of roustabout culture was marked by song and legend as the roustabouts added to the pageantry of steamboat culture. While the pageantry, bordered on ministrelization and served stereotypes held by whites regarding the marginalized rouster and steamboat culture, the songs offered another glimpse into the intersection between labor and the river. Further, not all songs were alike as the singing that marked the steamers' leave taking was a public event with crowds lining the levees to listen whereas other songs established a rhythm for the work of loading and unloading freight or feeding the boilers. The following song, though heard amid the ceremonial comings and goings of steamboats, still reveals an African American perspective of steamboat work and travel:

I's gwine frome de cotton fields, I's gwine from de cane
I's gwine from de ol log hut dat stan's down in de land;
De boat am in de river, dat comes to take me off,
An I's gwine to join de exodus, an' strike out fo' de no'f.

In this song, the mobility associated with steamboat work was memorialized. Other songs included implicit criticisms of the environment and working conditions in which African Americans labored.[25]

Still other roustabout songs such as the following accompanied the work of either loading or unloading cotton:

Come hyuh, you ole rouster!
Poke out yo' neck ve'y long!
Tell me which shoulder you wants it on!
Ole roustabout ain' got no home,
Makes his livin' by his shoulder bone![26]

Cotton, however, was not the only commodity being shipped on the river. One former riverboat pilot remembered roustabouts carrying kegs of nails weighing 107 pounds apiece for a distance of 200 yards. Each roustabout was expected to haul 75 kegs. In another account, sacks of corn weighing an estimated 150 pounds each were loaded on and off the steamers. In describing this work, the song "Carryin' Sacks" included the following lyrics:

"Carryin' Sacks"

I'm goin' up the rivuh to carry them sacks
I'm goin' up the rivuh to carry them sacks
I'm goin' up the rivuh to carry them sacks
I'll have yo' lap full uv dolluhs when I git back.[27]

In light of the hard work expected of the rousters, it is not surprising that other songs revealed the pathos of roustabout life:

> I'm wukin' my way back home,
> I'm wukin' my way back home,
> I'm wukin' my way back home, Baby
> I'm wukin' my way back home.
> Timber don't get too heavy fo' me,
> An' sacks too heavy to stack,
> All that I crave fo' many a long day,
> Is yo' lovin' when I git back.
> Oh, fireman, keep her rollin' for me,
> Let's make it to Memphis, Tennessee.
> Fo' my back is gittin' tired,
> An' my shoulder is gittin' sore.[28]

As the last song indicated, concomitant with the rich cultural production of the roustabout was a harsh everyday existence. In some firsthand accounts, the roustabouts were critical of their environment as one rouster, Ishe Webb, recalled quitting his job because of an environment of gambling and fighting. According to Webb, it was not unusual for fights to occur where someone would die, only to have their body tossed in the river. Gambling on board the steamboats was a constant as another roustabout, George Fortman, recalled a life where "we played cards, shot dice and talked to the girls who always met the boats." In still another memory by Ben Lawson, he claimed that as a roustabout, he "learned to gamble wid dice. I fought and gambled all up and down de Mississippi River." Similar to levee workers, roustabouts also remembered the mistreatment they received from their bosses. In one instance, however, Omelia Thomas recounted when her father was a roustabout, "They used to kick the roustabouts about and run them around but they never laid the weight of their hands on him." Included with the rough and tumble lifestyle, however, was the contention that working on the river, although hard, provided a good living. According to a nineteenth-century observer, wages for steamboat work could be as high as $60 a month.[29]

But images of an unconventional lifestyle dominated perceptions about the roustabout community. Late nineteenth-century observers of roustabouts who were especially critical of the roustabout's unconventional lifestyle disparaged lives that they perceived as dissolute and immoral. In one editorial published in the riverfront city of St. Louis, the author critiques their lives in the following manner: "His redeeming qualities are his cheerfulness under adversity, and his genius for singing rude melodies. All day long he saunters about the levee, or falls asleep in some nook where flies abound." Similar to the marginalized lives of other laboring classes, such as the barge haulers on the Volga, or loggers in lumber camps, the lives of the rousters became legendary, through song and folklore. When not loading or unloading the numerous steamships traveling up and down the river, roustabouts lived in dilapidated quarters, gaining reputations for frequenting taverns in the least desirable neighborhoods such as Natchez-Under-the-Hill. Songs such as the following reveal the uncertainty of the roustabout's life:

"Drunkard's Song"

I went down on the levee
Waitin' for the Miss Jaie Ray.
I knowed if I don't work fer the kind captain
I can't bring my Baby no pay.
While roamin' this wide world of sorrow
No cheerin' no comfort have I
And I think uv my ole whiskey bottle,
I know I'll drink till I die.[30]

Underlying the bravado, however, were sorrowful songs, such as the following known by its first line, "Ohio River, She's so Deep and Wide,"

Ohio River, She's so Deep and Wide
Lord, I can't see my poor gal
From the other side
I'm goin' to river, take my seat and sit down,
If the blues overtake me, I'll jump in the river and drown.
I've got the blues, I've got the blues,
Lord, I ain't got the heart to cry.[31]

Finally, the spirituality associated with the Mississippi River also found its way into the roustabout community. According to one former rouster, the following was often heard as the boat was leaving the levee:

"Red Sea"

When Moses was leadin' the Israelits, Red Sea,
Pharoah tried to ketch them jes' fo' spite, Red Sea,
Oh, Pharaoh he got drown-ded, drown-ded, drown-ded, Oh,
Pharoah he got drown-ded in the Red Sea.
I nevuh shall fo'git the day, Red Sea,
When Jesus washed my sins away, Red Sea.[32]

By the late 1880s, commercial steamboat travel was being replaced by railroads. Unable to control the river's alternating flooding and low water, railroads became a more reliable choice for transporting goods. For African Americans, steamboat work, similar to other river-related work, left a mixed legacy. For some, it provided a good living in comparison to other livelihoods but always coming at a cost (Figures 3.2–3.5).

FIGURE 3.2 Roustabouts.
Credit: Earl S. Miers River Photo Collection, Courtesy of the Tennessee State Library and Archives.

FIGURE 3.3 Hauling Bags in a Steamboat.
Credit: Earl S. Miers River Photo Collection, Courtesy of the Tennessee State Library and Archives.

FIGURE 3.4 Workers Roll Barrels.
Credit: Earl S. Miers River Photo Collection, Courtesy of the Tennessee State Library and Archives.

Although eclipsed commercially, by the early twentieth century, steam returned in the early twentieth century for recreational purposes as the old-fashioned paddle wheelers were launched as excursion vessels. The return of steam ushered in a new generation of African American music known as riverboat jazz. As African American musicians began to play in jazz bands on the excursion steamers, they continued the roustabout tradition associated with riverboat travel. Musical greats, such as Louis Armstrong, were part of an increasingly sophisticated riverboat jazz culture. Whether playing riverboat jazz or singing a roustabout tune, however, similarities between the two existed. For African Americans, laboring on the river offered mobility as twentieth-century jazz musicians viewed the excursion riverboats, operating on the Mississippi, Missouri, and Ohio Rivers, as opportunities to migrate North, leaving a Jim Crow South behind. For the white audiences located in riverfront towns, such as Memphis, St. Louis, New Orleans, or Dubuque, taking part in the riverboat excursions, the inclusion of riverboat jazz solidified the associations with steamboat travel and a false romanticized antebellum past. A nostalgia industry for the steamboat era and its surrounding levees was already in full swing, with works such as Mark Twain's *Life on the Mississippi*. Thus, for white excursion passengers, the music performed by African American musicians hearkened back to an earlier era whereas for African Americans, riverboat jazz signaled another escape—through the agency of the river—from an oppressive Southern white racist regime.[33]

Still another type of labor became common for African Americans living in the post-war era. Immediately after the Civil War, the prospects for owning

FIGURE 3.5 Assembling a Ramp.
Credit: Earl S. Miers River Photo Collection, Courtesy of the Tennessee State Library and Archives.

and farming their own land were promising. Although the proposed 40 acres and a mule that was initially promised to those formerly enslaved never materialized, Reconstruction lent optimism to life in the post-war South. By 1877, however, all this changed and by 1900, three-quarters of African Americans farming in the South were either sharecroppers or tenants. Despite the richness of the Mississippi Delta land in the Northwest Floodplain with its potential for substantial earnings derived from cotton crops, African American sharecroppers saw little profit. The system of sharecropping resulted in perpetual debt for

African American farmers. (One notable exception in the Delta region was the community of Mound Bayou.) Each year, money was borrowed to plant crops, usually at the company store, and even if the harvest was abundant, most of the earnings were used to offset the debt. Thus, for African Americans the end of the year accounting called "settling up" seldom resulted in profits. Critics referred to the system as one of debt peonage, with conditions deteriorating further when the Mississippi River unleashed one of its periodic, devastating floods.[34]

By the beginning of the twentieth century, the sharecropping system was routinely critiqued with abuses exposed in the African American journal, *The Crisis*. Founded by William DuBois in 1910, the journal reported on "inter-racial relations" in the twentieth-century United States, with particular concern regarding the current discriminatory practices against African Americans. Exposés such as the following appeared in the journal, alerting readers to the unfair farming arrangements in states such as Arkansas, Mississippi, and Louisiana. One Arkansan farmer, Steve Green, reported how he and other African American farmers were renting land from a Mr. Saddler in Crittenden County, Arkansas, which bordered the Mississippi River. In 1910, Saddler raised the rent from $5 per acre to $9, after which Green and other African American families moved from the property. After moving, the landlord informed Green that if he did not work for him, he could not work for anyone else in the county. So, when Green hired out to someone else in the county and Saddler, the former landlord, saw him, he started shooting at Green. According to Green's testimony, Saddler shot him in the arm, leg, and neck provoking Green to shoot Saddler, believing that he killed him. At this point, Green was on the run, escaping to the Mississippi River where he hid on an island in the river for three weeks. By the time, all of this was reported to *The Crisis*, he was being held and his friends were raising funds to hire a lawyer to defend him. In addition to accounts, such as Steve Green's, *The Crisis* also featured illustrations such as one entitled, "The Christmas Reckoning," which was another expose on the exploitative system of sharecropping exposé.[35]

In addition to the *The Crisis* editorials, other African American criticisms of the sharecropping system such as the following were recorded through the Federal Writers' Project. In the words of Henry Blake, who was born in Little Rock where his family farmed by the Arkansas River where his father also operated a skiff:

> After freedom, we worked on shares a while. Then we rented. When we worked on shares, we couldn't make nothing, just overalls and something to eat. Half went to the other man and you would destroy your half if you weren't careful. A man that didn't know how to count would always lose. He might lose anyhow. They didn't give no itemized statement. No, you just had to take their word. They never give you no details. They just say you owe so much. No matter how good account you kept, you had to go by their account and now, Brother, I'm tellin' you the truth about this. It's been that way for a long time. You had to take the white man's work on

> note, and everything. Anything you wanted, you could git if you were a good hand. You could git anything you wanted as long as you worked. If you didn't make no money, that's all right; they would advance you more. But you better not leave him, you better not try to leave and get caught. They'd keep you in debt. They were sharp. Christmas come, you could take up twenty dollar, in somethin' to eat and much as you wanted in whiskey. You could buy a gallon of whiskey. Anything that kept you a slave because he was always right and you were always wrong it there was difference. If there was an argument, he would get mad and there would be a shooting take place.[36]

In still another Federal Writers' Project interview, Lewis Brown succinctly summed up his father's experience as he farmed his whole life, raised all his children, "got wore out and pore."[37] Whether sharecropper or tenant, the Mississippi River was an ambivalent partner, the provider of rich farmland only to be realized by white planters and landlords.

With the conclusion of the Civil War in 1865, slavery in the United States was abolished. Yet, for African Americans living in the Southern states bordering the Mississippi River, the river remained a constant presence, possessing many of the same associations that had existed since the early 1700s. Even with the end of a slave regime, the Mississippi River—through a post-war history of forced labor, sharecropping, and bulldozing—represented a site of violence and trauma, while fostering a sense of escape through migration combined with a spirituality and refuge as seen through song and the experiences of groups such as the Exodusters. For many African Americans who remained in the South, the post-war years introduced new atrocities as white Southerners sought to retain their positions of privilege, resting in part upon a cheap source of labor. Through debt-ridden systems of sharecropping and tenancy, African Americans who farmed in the states bordering the Mississippi River faced lives of peonage despite the rich farmlands they worked on. Yet, within this history fraught with violence is a story of continuing resistance chronicled through a rich cultural outpouring found in song, prose, folklore, and oral histories.[38]

In contrast with this narrative of the river, white Southerners were crafting a New South, mythologizing a past with manicured plantations resting upon a contented slave regime. The Mississippi River figured prominently in the portrayal, illustrated by nostalgic depictions of plantation-era homes lining the riverbanks with African Americans tending acres of cotton fields. For example, in 1872, *Picturesque America* was published with its antique "Engraving of a Plantation House on the Mississippi River." Another celebratory nod to the past came from the late nineteenth-century prominent Louisiana historian, Alcee Fortier. In his text, he described New Year's Day on the plantation, summoning a bygone, harmonious world when he wrote the following:

> New Year's Day on the plantation was an occasion of great merriment and pleasure for the slaves. . . . At daylight, on the 1st of January, the rejoicing

> began on the plantation; everything was in an uproar, and all the negroes, young and old, were running about, shaking hands and exchanging wishes for the new year.[39]

This wistfulness for the antebellum past was also expressed when discussing the end of the steamboat era.[40]

Concomitant with the narrative was another one of an emergent New South as land companies, such as the Southern Alluvial Land Association, formed in 1916, promoted the Delta lands and a welcoming Southern community, to recruit those interested in farming. Resembling the booster environment found in land companies throughout the Western United States in the late nineteenth and early twentieth centuries, the Southern Alluvial Land Association promised profitable returns on agricultural land with minimum effort. Complementing their narrative was a tamed Mississippi River as their brochure boasted of a "chain of titanic levees that conquered the Mississippi River." In their literature, no mention was made of the large African American population that inhabited the state although the photographs of the area showed African Americans performing much of the labor. The coexistence of the two narratives persisted into the twentieth century.[41]

Despite the Southern Alluvial Land Association's boast of a "chain of titanic levees," reflecting the U.S. Corps of Engineers' ongoing efforts to subdue the river, flooding still occurred and levees failed. In the words of William Stone, who was born into slavery and after the Civil War worked on the Mississippi bottom:

> Floods come down, no matter what time of year. One day Old Man River be runnin' 'long, jes' as peaceful and quiet, and everybody happy. Everybody meet de boats at de landin'. Den way in de night you wake up and hear a roarin' like thunder and dat river be on a tear. Folks know he am in de ugly mood, and starts movin' to higher ground. Everybody what have a wagon and mule gits out. Some jes' gits to de levee. . . . Old Man River sho' treach'ous. After he go on one he rarin' and tearin' spells, den he gwine be so peaceful and quiet like. Look like he try to make up he meanness.

However, Stone, like Henry Bib in Chapter 2, recognized the multidimensionality of the river. While he realized the disruption that an angry Mississippi River caused, he also praised the river's beauty when he stated, "old Man River was sho' purty in de fall, when dem wild geese come in droves and de blossoms red and yaller." Finally, in another poetic turn when discussing the river, Stone portrays a river with agency when referencing levee building to hold "Old Man River back when he start prowlin' round 'gain." Stone's predictions were accurate as the Lower Mississippi River flooded in 1891, 1892, and 1893, inundating homes and destroying the livelihoods of thousands "as it started prowlin' round".[42]

Although both black and white Southerners living along the river suffered when the river overflowed its banks, the treatment of each differed. In 1912, for example, when another flood occurred with the Torres crevasse, *The Crisis* observed that it had been reported in the State of Louisiana: "Negro flood refugees are being compelled to work on the levees." Upon hearing the report, the National Association for the Advancement of Colored People (NAACP) wrote to Louisiana's governor for verification. While the governor denied treating African Americans differently than others affected by the flood, *The Crisis* printed responses from other African Americans disputing the governor's claim. One response claimed that there were about 6,000 African Americans camping in the city of Baton Rouge. In citing their conditions, the observer noted:

> in many cases [African Americans were] without enough clothes on to present themselves in public, the shame and suffering of the refugees cannot be described. . . . Hundreds are together in buildings with nothing to lie upon but the bare floors and nothing to cover with but the scanty clothes they are in all day. . . . their fathers are away working for their rations on the levees . . . .

The same writer contrasts this with the treatment for the white refugees housed in the "best places obtainable for their comforts," and the work they are assigned is "overseeing the working Negroes and acting as guards on the levees. . . ." The contrast in treatment for flood victims persisted throughout the twentieth century, reaching its nadir in 1927 with the Great Flood. Despite an outcry from African Americans, distinctions were made between white and black flood victims as recently as Hurricane Katrina in the twenty-first century. The distinctions produced different narratives where the levees and other Mississippi River "improvements" were assigned differing roles, as the next chapter will reveal.[43]

By the end of the nineteenth century, African Americans experienced two eras—the heyday of the Mississippi Valley's cotton empire realized through their enslavement and the end of slavery, only to be replaced by the punitive, Jim Crow South. The Mississippi River figured prominently in both eras. In the antebellum period, the river intersected with African American lives on several levels, acting as a liberator and a means to enslavement, inculcating not only memories of violence and trauma, but also empowerment. After the war, a new chapter was being written for African Americans and the Mississippi. Now the river represented not only escape but also brutality as the site of a growing number of levee camps. While the levees were built by slaves in earlier periods, the convict lease system of building public works ushered in new expectations by white Southerners unencumbered by a slave regime. Just as the river became increasingly commodified, so did African American lives building and maintaining levees. Yet the levees were also platforms for escape—whether through the Exodusters' use of levees as encampment sites or sites for flood victims, fleeing the river's rise. Thus, Mississippi River levees became a constant in the collective memory of African Americans.

During the same period, another chapter ended for the river: commercial steamboat travel. While American commentators and novelists, such as Mark Twain, lamented the end of steam, African Americans' experience with the river through steamboats reflected a portrayal shaped by hard labor. Their songs and stories revealed a kinship with the river derived from their labor, whether roustabout or steward. Photographs from the era document the labor exacted from African Americans working on the steamships. The twentieth century, however, introduced new experiences with the river, which would be captured through song, memoirs, and journalistic reports. While the commitment to modernize the river, on the part of engineers, planters, and politicians, marked a departure from earlier river relationships in myriad ways, the river of 1719, when Africans first arrived in New Orleans, remained. For the next 100 years, the river persisted in delivering unpredictable flooding, destroying homes, and livelihoods. But the same river remained the highway north, facilitating the Great Migration in the twentieth century while also a conduit for a cultural outpouring that included musical innovations, such as the blues, not to mention a rich literary presence. Thus, for African Americans, the river continued to play multidimensional roles in their lives.

Notes

1 *St. Louis Globe Democrat*, 13 March 1879; 17 March 1879; 25 April 1879. Lee Ella Blake, "The Great Exodus of 1879 and 1880 to Kansas" MA Thesis, (Kansas State College, 1942):1–2 accessed at https://core.ac.uk/download/pdf/14342729.pdfThe classic work on the Exodusters remains Nell Irvin Painter, *Exodusters. Black Migration to Kansas after Reconstruction* (New York: W.W. Norton & Company, 1976). Painter was the first historian to record the story of the Exodusters in its entirety, chronicling another chapter in African American history that reveals African American self-determination to resist the tyranny of the post–Reconstruction era. For an online overview of the Exodusters, see Samantha Gibson. Exodusters: African American Migration to the Great Plains. 2018. Retrieved from the Digital Public Library of America, http://dp.la/primary-source-sets/exodusters-african-american-migration-to-the-great-plains

2 *New York Times*, 22 May 1879, p. 1.

3 John Solomon Lewis, Leavenworth, Kansas, 10 June 1879, Boston *Traveller in New Orleans Southwester Christian Advocate*, 3 July 1879 quoted in Painter, *Exodusters*, p. 4.

4 At the time of the Exoduster movement, many Exodusters believed they would receive assistance from the federal government once they reached Kansas. See Painter, *Exodusters*, pp. 177–179; *The Times* (London), 23 April 1879, p. 5; St. Louis Globe Democrat, 17 March 1879, p. 1.

5 Blake, "The Great Exodus of 1879 and 1880 to Kansas," pp. 42–44.

6 *Report and Testimony of Select Committee to Investigate the Causes of the Removal of the Negroes From the Southern States*, U.S Senate, Executive Document no. 693, 46th Con gress 2nd Session (Washington, 1880):xvii (Hereafter referred to as *Senate Report 693*); Carl Schurz, *Report on the Condition of the South*, U.S. Senate, Executive Document no. 2, 39th Congress 2nd session, (Washington, 1865) accessed at https://www.gutenberg.org/cache/epub/8872/pg8872.html; W.E.B. DuBois, *Black Reconstruction in America: An Essay Toward a History of the Part Which Black Folk Played in the Attempt to Reconstruct Democracy in America, 1860–1880* (1935, rpt. New York: Free Press, 1997):vii.

7 General H. Sheridan, *Senate Report 693*:xviii. Scholarship on the Reconstruction Era is voluminous but one of the classics remains W.E.B. DuBois, *Black Reconstruction in America, 1860–1880* (Cambridge: Oxford University Press, 1935; rpt. 2007). Another oft-cited work is Eric Foner, *Reconstruction, Updated Edition: America's Unfinished Revolution, 1863–1877* (New York: Harper Perennial Modern Classics, rev. ed., 2014). For an overview of Reconstruction historiography, see John David Smith, ed., *Reconstruction: Interpreting American History* (Kent: Kent State University Press, 2016). After Reconstruction, lynchings were an all too-common experience in the South. In documenting these dire conditions, in the late nineteenth and early twentieth centuries, *The Crisis* cited the number of lynchings of African Americans from 1885 onward. For example, in Volume 2, the number of lynchings had reached an estimated 2,425 since 1885. (Vol. 2, 1910–1912, p. 26).

8 Henry Adams, *Senate Report 693*:27; *Senate Report 693*:xxiii. In analyzing the violence reported in accounts, such as Sheridan's and Adams', historians initially portrayed "bulldozers" as poor, Southern whites acting independently. More recent studies, however, contend that the upper-class white Southern community tacitly approved as acts of bulldozing would not have occurred without their acquiescence. One of the best texts on life in the South for African Americans after the Civil War, particularly with the end of the Reconstruction, remains W.E. DuBois, *Black Reconstruction in America, 1860–1880* (New York: Free Press, 1935; rpt. 1997). In assessing this period, DuBois observed: "In all cases, there was plain and indisputable attempt on the part of the Southern states to make Negroes slaves in everything but name." See DuBois, *Black Reconstruction*, p. 167. Although dated, another classic study on the post–Civil War South is C. Vann Woodward, *Origins of the New South: 1877–1913* (Baton Rouge: Louisiana State University Press, 1971). Woodward analyzes the development of Jim Crow and the changing economic conditions for African Americans at the conclusion of the Reconstruction.

9 Bryan M. Jack, *The St. Louis African American Community and the Exodusters* (St. Louis: University of Missouri Press, 2008):26. In Jack's text, he offers an excellent overview of the Exodusters' experiences once they reached St. Louis, including the work of the St. Louis African American community in providing assistance to the refugees.

10 One of the most frequently photographed images of the Exodusters captures groups waiting on the levees whether for steamships to take them to the "Promised Land" or encampments outside St. Louis. The levee as a makeshift campsite will persist into the twentieth century with environmental disasters such as the 1927 Flood.

11 "The Land That Gives Birth to Freedom," Kansas Memory, Kansas Historical Society accessed at https://www.kshs.org/index.php?url=km/items/view/218451. The role of "Pap" Singleton in the Exoduster movement reveals a millenarian strain in the Exoduster movement that has been debated by historians.

12 Jack, *The St. Louis African American Community*, p. 19. During the 1880s, agriculture did well in Kansas and Nicodemus prospered. However, with the Depression of the 1890s and two major railroads choosing routes that bypassed the town, Nicodemus' economy began to suffer. However, the community did survive until World War I and then declined again during the interwar years. See Nicodemus, Graham County, Kansas Historical Society at https://www.kshs.org/kansapedia/nicodemus-graham-county/12157

13 In the Yazoo Delta region of the Mississippi River Basin, the construction of levees was essential for development. The area, though rich in alluvial soil, required the means to clear the thick vegetation surrounding the river that also invited disease. Flooding was always a threat and thus for planters in this region, federal financing of levees became a priority after the war. For an excellent discussion of the region and the efforts to attract federal funding for levee construction, see James C. Cobb, *The Most Southern Place on Earth: the Mississippi Delta and the Roots of Regional Identity* (New York: Oxford University Press, 1992); Edward L. Ayers, *The Promise of the New South: Life After Reconstruction* (New York: Oxford University Press, 1992).

14 The contested 1878 presidential election was instrumental in the allocation of federal funds to the South. Southern Democrats supported Hayes election with the understanding that federal resources would be awarded to the South for internal improvements such as the levee program. See Todd Shallat, *Structures in the Stream: Water, Science, and the Rise of the U.S. Army Corps of Engineers* (Austin: University of Texas Press, 1994):193. For a brief history of the Mississippi River Commission, see U.S. Army Corps of Engineers, *Mississippi River Commission History* (June 2018) accessed at https://www.mvd.usace.army.mil/Portals/52/docs/MRC/11_MRC_History_WEB.pdf

15 Tabitha LaFlouria, ""Under the Sting of the Lash": Gendered Violence, Terror, and Resistance in the South's Convict Camps" *Journal of African American History*, Vol. 100, No. 3 (Summer 2015):366–384. LaFlouria lists some of the more petty offenses that resulted in imprisonment. She also demonstrates how Southerners used imprisonment as one means to enforce a white post-war hierarchy. For further discussion on the linkage between state building and the penal system in Louisiana, see Christina Pruett Hermann, "Specters of Freedom: Forced Labor, Social Struggle, and the Louisiana Penitentiary System, 1835–1935" Ph.D. Dissertation (Michigan State University, 2015) accessed at https://d.lib.msu.edu/islandora/search/program_ss:(%22History%22)+AND+subject_display:(%22Prisons%22)?&sort=fgs_label_s+asc Hermann challenges earlier works associating the lack of penal reform in the South to "backwardness." Instead, she sees the absence of reform and the development of a convict labor force as the product of "advanced state-building, planter power and infrastructural development." She also contends that the post-war prison regime was a reaction to the resistance of former slaves and poor whites who sought to shape and define their labor. Hermann, "Specters of Freedom," pp. 3, 239. Adding to the argument that the nineteenth-century penal system was "an auxiliary arm of capitalist industry and commerce," is Steve Fraser and Joshua B. Freeman, "In the Rearview Mirror: Barbarism and Progress, the Story of Convict Labor" *New Labor Forum*, Vol. 21, No. 3 (Fall 2017):96. Unfortunately, the statistics for imprisonment of African American males in the twenty-first century have not improved. According to Jed Rakoff, "840,000 or nearly forty per cent of today's prisoners in the U.S. are African American males." See Rakoff, *Why the Innocent Plead Guilty and the Guilty Go Free* (New York: Farrar, Straus & Giroux, 2021), Kindle ed., n.p.

16 John C. Willis, *Forgotten Time: The Yazoo-Mississippi Delta after the Civil War* (Charlottesville: University Press of Virginia, 2000):2.

17 Nathan Cardon, "Less than Mayhem": Louisiana's Convict Lease, 1865–1901" *Louisiana History*, Vol. 58, No. 4 (Fall 2017):421–422; *New Orleans Daily Picayune*, 22 March 1886, p. 2. Most of the scholarship on the convict lease system examines its role in the modernization of the post-war South. In modernization discussions, the development of Jim Crow and the South's efforts to link race and criminality are included. But examining the convict lease system from the perspective of African Americans living along the Mississippi River is not a focus except in judgments comparing the convict lease system and slavery. For a few notable works reviewing the convict lease system in the context of the New South, consult Mark T. Carleton, *Politics and Punishment: The History of the Louisiana State Penal System* (Baton Rouge: Louisiana State University Press, 1971); Steve Fraser and Joshua B. Freeman, "In the Rearview Mirror: Barbarism and Progress: The Story of Convict Labor" *New Labor Forum*, Vol. 13, No. 1 (Fall 2017):94–98; LaFlouria, *Chained in Silence: Black Women and Convict Labor in the New South* (Chapel Hill: University of North Carolina Press, 2015).

18 Carleton, *Politics and Punishment*, p. 23; *New Orleans Daily Picayune*, 30 June 1884, p. 4; Matthew J. Mancini, *"One Dies, Get Another": Convict Leasing in the American South, 1866–1928* (Columbia: University of South Carolina Press, 1996):67; Hermann, "Specters of Freedom," p. 270. Mancini includes more statistics demonstrating the brutality of the penal system in Louisiana and Mississippi. For example, in 1887, the mortality rate for convicts in Mississippi was 16%, which, undoubtedly, contributed to another statistic in which the number of prisoner escapes from 1876 to

1899 in Mississippi was 1715. Out of this number, 735 were recaptured. See Mancini, "*One Dies, Get Another*," pp. 67–68. For an overview of critiques and exposés, see Vernon Lane Wharton, *The Negro in Mississippi, 1865–1890* (New York: Harper and Row, 1947):241. Several oral histories from the 1930s to the 1960s by those who lived at the Angola Penitentiary, both employees and prisoners, were compiled by Anne Butler and C. Murray Henderson. See *Angola, Louisiana State Penitentiary: A Half-Century of Rage and Reform* (Lafayette: University of Louisiana at Lafayette Press, 1990). One of the stories includes the attempted escape by James Bruce, a prisoner, by way of the Mississippi River. (Angola covered 18,000 acres of Mississippi River bottomland.) The current, however, was too strong and the prisoner drowned. His escape, however, from Angola baffled prison employees as he changed his appearance by wearing women's clothing. Prison officials looked for him for almost two weeks before finding his body. *Angola*, pp. 34–56. For a discussion of Mississippi's equivalent to Angola, see David M. Oshinsky, *"Worse Than Slavery": Parchman Farm and the Ordeal of Jim Crow Justice* (New York: Free Press, 1996).

19 Hastings H. Hart, "Prison Conditions in the South" *Proceedings of National Prison Association*, 1919:200 as quoted in Matthew J. Mancini, *One Dies, Get Another,* p. 2; George W. Cable, *Silent South: Together with the Freedmen's Case in Equity and the Convict Lease System* (1885, Antique Reprints, 2016):171. In a review of the convict lease system in Louisiana, "prison conditions declined dramatically during James' period of 1870–1884." The same report said that James sent most of his prisoners to work on levees. See U.S. General Services Administration, *Hard Labor: History and Archaeology at the Old Louisiana State Penitentiary* (Baton Rouge, 1991):11.

20 For a discussion of everyday life in the levee camps, see Richard M. Mizelle, Jr., *Backwater Blues: The Mississippi Flood of 1927 in the African American Imagination* (Minneapolis: University of Minnesota Press):129–136.; J. Crowley, "Shack Bullies and Levee Contractors: Bluesmen as Ethnographers" *Journal of Folklore Research*, Vol. 28, No. 2 (1991):156.

21 Alan Lomax, *The Land Where the Blues Began* (New York: The New Press, 1993):242. When discussing the conditions in the levee camps, Alan Lomax is also a source for firsthand accounts. However, his collecting has come under criticism for his bias toward blues music, omitting other genres. Also, critics have cited his focus upon the prison population and their reluctance to share their experiences and knowledge with an outsider. Still, most scholars would agree that his interviews revealed valuable insights into the genre as well as the experiences that contributed to blues music. See Richard Paul, producer, "In the Field of Folk Music, Alan Lomax is a Giant—If a Flawed and Controversial One" *The World* (February 2015) accessed at https://www.pri.org/stories/2015-02-10/field-folk-music-alan-lomax-giant-if-flawed-and-controversial-one

22 Washboard Sam, "Levee Camp Blues" (1941) accessed at https://genius.com/Washboard-sam-levee-camp-blues-lyrics

23 Lucille Bogan, "Levee Blues" (1927) accessed at https://www.discogs.com/Lucille-Bogan-Levee-Blues-Sweet-Patunia/release/14027021

24 For further discussion on the aftermath of the 1927 Flood, see Chapter 4.

25 Frank Donovan, *River Boats of America: From Flatboats to Floating Palaces, From the Cotton Blossom to the Staten Island Ferry* (New York: Thomas Y. Crowell Company, 1966):225–226; Herbert Quick and Edward Quick, *Mississippi Steamboatin': A History of Steamboating on the Mississippi and Its Tributaries* (New York: Henry Holt and Company, 1926):250. For further discussion of the criticisms implied or "coded" within the rousters' songs, see Thomas Buchanan, *Black Life on the Mississippi* (2004).

26 Quick, *Mississippi Steamboatin'*, p. 250.

27 Mary Wheeler, *Steamboatin' Days: Folk Songs of the River Packet Era* (Baton Rouge: Louisiana State University Press, 1944):29–30.

28 Donovan, *River Boats of America*, p. 226.

29 Ishe Webb Interview, Federal Writers' Project: Slave Narrative Project, Vol. 2, Part 7, Arkansas, (nd):Image 80 accessed at https://www.loc.gov/resource/mesn.027/?sp=85.

In Webb's interview, although enslaved he comments upon his parentage in that his mother was Native American and his father was white. George Fortman Interview, Federal Writers' Project: Slave Narrative Project, Vol. 5, Indiana (nd):Image 88 accessed at https://www.loc.gov/resource/mesn.050/?sp=88; Ben Lawson Interview, Federal Writers' Project: Slave Narrative Project, Vol. 13, Oklahoma, 5 November 1937: Image 182 accessed at https://www.loc.gov/resource/mesn.130/?sp=181; Omelia Thomas Interview, Federal Writers' Project: Slave Narrative Project, Vol. 2, Part 6, Arkansas, (nd): Image 305 accessed at https://www.loc.gov/resource/mesn.026/?sp=305; Joe Mayers Interview, Federal Writers' Project: Slave Narrative Project, Vol. 2, Part 5, Arkansas (nd): Image 74 accessed at https://www.loc.gov/resource/mesn.025/?sp=73; Fortman, Federal Writers' Project, Image 88; Lawson, Federal Writers' Project, Image 182; J. T. Trowbridge, *The South* (Hartford: L. Stebbins, 1866):388–389.

30 "The Levee and Its Fragrant People," *Republic* (St. Louis), 5 August 1875, accessed at Ruth Ferris Collection of River Life and Lore, General Box 22, Herman T. Pott National Inland Waterways Library, St. Louis Mercantile Library, University of Missouri-St. Louis Ruth Ferris Collection of River Life and Lore, General Box 22. Another resource for roustabout songs can be found in the Mary Wheeler Collection, McCracken County Public Library, Digital Collections accessed at https://digitalcollections.mclib.net/luna/servlet/detail/McCracken~13~13~129~2241:Drunkard-s-Song?sort=title%2Csubject%2Cdate%2Ccoverage&qvq=sort:title%2Csubject%2Cdate%2Ccoverage;lc:McCracken~13~13&mi=62&trs=374

31 Wheeler, *Steamboatin' Days*, pp. 82–83.

32 Wheeler, *Steamboatin' Days*, pp. 70–71.

33 For an excellent discussion of the importance of riverboat jazz and its importance for African Americans living in the Lower Mississippi Valley, see William Howland Kenney, *Jazz on the River* (Chicago: University of Chicago Press, 2005). To learn more about Mississippi River representations in the late nineteenth century, see the impressive work of Thomas Ruys Smith, *Deep Water: The Mississippi River in the Age of Mark Twain* (Baton Rouge. Louisiana State University Press, 2019.)

34 Seth C. McKee, "Politics in Black and White: The Mississippi Delta" *Defining the Delta: Multidisciplinary Perspectives on the Lower Mississippi River Delta*, ed. Janelle Collins (Fayetteville: University of Arkansas Press, 2015):133. In McKee's article, he discussed the parallel of the richness of the Delta lands with the largest number of African Americans farming the land in the "most racially oppressive state." p. 133. Thus, for African Americans who wanted to farm, sharecropping and tenancy were the most viable options as land ownership in the South was rare. By 1900, 75.3% of African American farmers in the South were either sharecroppers or tenants. See Woodward, *Origins of the New South*, p. 206.

35 *The Crisis*, Vol. 1 (1910–1911):14; Vol. 2 (1910–1912):19. In another *Crisis* article, it was reported that Mississippi planters in the fertile Black Belt lands were "warning labor agents not to entice their colored laborers away to Arkansas." P. 7, Vol, 4, 1911–1912.

36 Henry Blake Interview, Federal Writers' Project. Slave Narrative Project, Vol. 2, Part 1, Arkansas (nd):Image 183 accessed at https://www.loc.gov/resource/mesn.021/?sp=180&st=text

37 Lewis Brown Interview, Federal Writers project, *Arkansas Slave Narratives* (December 1938; rpt. 2007):Image 297.

38 In Hermann's dissertation, she argued that the strict post–Civil War laws curtailing African Americans' civil rights were, in part, a product of African American and "some poor white resistance: to work on their own terms." "Specters of Freedom," p. 239.

39 William Cullen Bryant, ed., *Picturesque America* (New York: D. Appleton & Co., 1872, 1874); Alcee Fortier, *Louisiana Studies: Literature, Customs and Dialects, History and Education* (New Orleans: F.F. Hansell and Brother, 1894) as quoted in Rosan Augusta Jordan and Frank De Caro, "In this Folk-Lore Land": Race, Class, Identity, and Folklore Studies in Louisiana" *The Journal of American Folklore*, Vol. 109, No. 431 (Winter, 1996):42.

40 For an excellent overview regarding the end of the steamboat era and the nostalgia surrounding the post–Civil War Mississippi River, see Thomas Ruys Smith, ""The Mississippi Was a Virgin Field": Reconstructing the River before Mark Twain, 1865–1875" *Mark Twain Journal*, Vol. 53, No. 2 (Fall 2015):24–66.

41 Southern Alluvian Land Association, *The Call of the Alluvial Empire: Containing Authentic Information About the Alluvial Region of the Lower Mississippi Valley, Particularly the States of Arkansas, Tennessee, Mississippi and Louisiana* (Southern Alluvian Land Association, 1919):5 accessed at https://play.google.com/books/reader?id=e8VB-AQAAMAAJ&pg=GBS.PP2&hl=en. In true booster fashion, the pamphlet goes on to tout ". . .land more fertile than Egypt's far-famed Nile; a New Eden in the heart of America," p. 5. For an insightful discussion of the association, see Nancy Elizabeth Woodruff, *American Congo: The African American Freedom Struggle in the Delta* (Cambridge: Harvard University Press, 2003).

42 William Stone Interview, Federal Writers' Project: Slave Narrative Project, Vol. 16, Part 4, Texas (nd):Image 72 accessed at https://www.loc.gov/resource/mesn.164/?sp=70&st=text; Woodward, *Origins of the New South*, p. 270.

43 *The Crisis*, Vol. 4, No. 3 (July 1912):127–128.

4

THE GREAT FLOOD OF 1927

A Modern River

FIGURE 4.1 Camping on the Levee, Mississippi Floodwaters.
Credit: Courtesy of Memphis and Shelby County Room, Memphis Public Libraries

In 1927, Charley Patton captured the drama and misery of the Great Flood of 1927 for African Americans living in the Yazoo Delta with the following lyrics (Figure 4.1):

> *"High Water Everywhere"*
>
> The back water done rolled lord, and tumbled, drove me down the line
> The back water done rolled and tumbled, drove poor Charley down the line
> Lord, I'll tell the world the water done struck Drew's town

DOI: 10.4324/9781315617077-5

Lord the whole round country, lord creek water is overflowed
Lord the whole round country, man, is overflowed
(spoken: you know, I can't stay here, I'm bound to go where it's high boy.)
I would go to the hill country, but they got me barred

....

Looky here, the water dug out, Lordy, levee broke, rolled most everywhere
The water at Greenville and Leland, Lord, it done rose everywhere
(spoken: boy, you can't never stay here.)
I would go down to Rosedale, but they tell me there's water there.

Back water at Blytheville, backed up all around
Back water at Blytheville, done struck Joiner town
It was fifty families and children. Tough luck, they can drown

The water was rising up in my friend's door
The water was risins up in my friend's door
The man said his womenfold, "Lord we'd better go"

Oh Lordy, women is groaning down
Oh Lordy, women and children sinking down
(spoken: Lord have mercy)
I couldn't see nobody home, and was no one to be found.[1]

These lyrics framed familiar themes for African Americans living along the Mississippi River—loss and displacement intertwined with the possibility of violence. In the lines "barred from the hills," Patton referenced the actions by white planters and others, forbidding African Americans to leave the levees. The prohibition against leaving served two purposes: secured labor to maintain the levees and labor for planting and harvesting crops. Perhaps the most powerful optic from Patton's song, however, was the imagery of water everywhere as those living on the bottomlands saw high water swamp their homes, forcing them to flee, ruining their chances to plant spring crops, and destroying livelihoods. Born in the 1890s, Patton spent much of his life on the Dockery Plantation, located by the Sunflower River, part of the Mississippi River Basin. He knew the Delta landscape and sang with authority, communicating an individual perspective as well as that of other African Americans living along the river. As a result, "High Water Everywhere" was one of the most popular songs about the 1927 Flood.[2]

But there were others who immortalized the 1927 Flood through a new blues genre known as the flood blues. In Barbecue Bob's 1927 song, "Heavy Water Blues," he lamented:

I was walking down the Levee with my head hanging low
Looking for my sweet mama but she ain't here no more
That's why I'm crying Mississippi Heavy Water Blues
Lord, Lord, Lord, I'm so blue, my house got washed away

And I'm crying "how long for another pay day"
That's why I'm crying Mississippi Heavy Water Blues
I'm sitting here looking at all of this mud
And my girl got washed away in that Mississippi flood.[3]

In these lyrics, Barbecue Bob, reinforcing Patton's sentiments, acknowledged the hopelessness and despair that many experienced as they mourned the loss of loved ones as well as property. For African Americans living along the Mississippi bottomlands, the Great Flood of 1927 was particularly devastating as their losses were much greater because their homes were the first to be flooded, initiating one of the first major migrations north.

Once called the "greatest peacetime disaster in the United States," the 1927 Mississippi River Flood is the subject of numerous texts, ranging from engineering explanations regarding the failure of a levees-only policy to political accounts of leaders, such as Herbert Hoover, in times of crisis. Only in recent years, however, have new histories emerged chronicling the flood and its impact on the lives of African Americans, who comprised the majority of flood refugees. Warnings of a potential flood began with excessive rain in August 1926 in the Upper Mississippi Valley, which started again in January 1927 in the Lower Mississippi Valley, straining already weakened levees. Early snowmelt contributed to rising river levels, compounded by heavy rains in April; making conditions ideal for severe flooding.. Adding to the excessive rainfall were three tornadoes in mid-March, further weakening levees in the lower valley. Compounding the weather conditions was the loss of forest cover and wetlands to absorb and slow the floodwaters, underlying the fact that this was not a natural disaster. Worsening the situation was a levee system built on the erroneous assumption that levees could contain the river. Despite the contributing factors, reports consistently laid blame solely on the weather. The consequences of the catastrophic flood are revealed through a litany of daunting statistics that begin with an official fatality count of 250, although later scholars estimated that the death toll could have reached 1,000. Of those 1,000, at least 100 African American men drowned in a single night as National Guard troops forced them to remain at the Mounds Bayou levee in a last-ditch effort to save the levee. This levee, outside of Greenville, Mississippi, along with more than 200 others, failed with floodwaters covering an area 70 miles wide, affecting seven states, inundating more than 26,000 square miles, leaving almost 700,000 without homes and property damage reaching the billion-dollar mark. The floodwaters did not completely recede until the end of the 1927 summer. Of those displaced, the majority were African Americans (approximately 555,000), revealing a world where Jim Crow laws prevailed and peonage was the norm on the last remaining Southern cotton plantations. In the midst of multiple flood narratives, the experiences of African Americans living in the Yazoo-Mississippi Delta are retold, emphasizing their harsh, often brutal mistreatment by Southern whites—witnessed by indifferent federal and aid officials—resulting in a mass migration to urban areas in the North and South.[4]

Shifting focus from the Mississippi River as a whole, to one region, the Yazoo-Mississippi Delta, the area most devastated by the 1927 Flood, many of the same themes characterizing African Americans' earlier history with the river emerge. The Delta occupied a swath of rich alluvial land extending from Memphis in the north to Vicksburg in the south and remained largely uncultivated until the late nineteenth century. With an environment that deterred settlement, one early resident described her first impressions of the land as possessing "the thickest timber [she] had ever seen. Oak, gum, ash, hackberry, and poplar stood so thick, with no underbrush, only big blue cane growing rank and tall, almost to the limbs of the trees." And as she came nearer to the river, "the ground was sandy, but it was black sand, and the woods were thinner; there were fewer trees but larger-big old cottonwoods and sycamores that seemed . . . like their tops were lost in the sky." Not surprisingly, one of the first industries in the area was lumbering as the land was cleared for planting cotton. After deforesting the land, the work of levee building as discussed in Chapter 3 began in order to protect these rich plantation lands from Mississippi River floodwaters. For African Americans not sharecropping in the post–Civil War South, the back-breaking levee work was another livelihood. (However, many African Americans, hearing of the rich alluvial soil on the Delta lands, also moved to the Delta lands, entering into sharecropping arrangements.) By the time of the 1927 Flood, levees extended over 1,000 miles from Cairo to the Gulf of Mexico.[5]

After the Civil War and the floods that occurred in the late nineteenth-century, with major floods in 1882, 1884, 1890, 1897, 1903, 1912, 1913, and 1922, building levees became the mantra for flood control in the late nineteenth and early twentieth centuries as engineers saw their construction as the surest way to contain the unpredictable Mississippi. With the formation of the Mississippi River Commission, flood control and later navigation, under the leadership of the U.S. Army Corps of Engineers, was initiated guided by the belief that by corseting the river with levees, a scouring action would result and instead of the river overflowing its banks, the current would deepen. This belief that levees could dictate the river's behavior remained ingrained in the engineering community throughout the twentieth century. Of course, the 1927 Flood challenged this idea as floodwaters broke through more than 200 levees, covering an area 70 miles wide in some areas. Still, after the floodwaters receded, politicians and planters began the familiar refrain that what was needed was "more and better levees."[6]

Returning to the days surrounding the flood, not all breaks were equal and towns such as Greenville, Mississippi suffered more than others. According to a local newspaper, the *Daily Democrat Times*, the levee protecting Greenville at Mound Bayou broke at 7:45 a.m. on April 21, 1927 despite the efforts of "fifteen hundred men battling all night in terrific rain against raging torrents." Instead of blaming the levee break on a misplaced faith in the ability of levees to prevent flooding, on April 21, responsibility for the levee break was attributed to the weather as a Greenville reporter noted that

> never in the history of the Mississippi River has such a terrific flood swooped down on the river as this morning. Never before had the waters at such high stage made such rapid rises. During the past 24 hours a rise of one foot was recorded, which is without precedent.

According to one flood survivor, the water in Greenville was "as much as ten and twelve feet deep within the city" and "it took a little time for the water to go back down." Other recollections describe the flood beginning in Greenville as a "little silver ribbon of water coming down the gutters on either side" but in a few hours

> the streets were covered and then you could see huge oil tanks from Standard Oil Company on North Broadway come tumbling down, rolling down. You could see ice boxes, all sorts of household wares coming down. Now, when they got into the intersection of Broadway and Nelson, the current was very, very strong and they would just go down and you wouldn't see them anymore, some of the smaller articles. There was a dairy, Blocker Dairy, and the cows came down. Now, that's where the noise was. They were lowing most pitifully, and the little calves were trying to keep up with their mothers. Many a calf just went down with that current, right there!

Another Greenville memory that encompassed the sounds of the flood recalled, "I hear'd a noise like de wind an' I asked. . . Is dat a storm? Dey said, "no, dat's de rivercomin' th'ought an' you better come back 'fore de water ketch you." The response by Greenville city officials was immediate and from the outset they differentiated between African American and white residents.[7]

In the evacuation plan outlined by Greenville political leader Leroy Percy, who headed the citizens relief committee, he reported, "We are urging all white women and children to leave the city." He went on to say, "White men may also go although there is need for them to stay and we are assured they will. There is need for Negro men to stay in order to establish the camp." According to another report, Percy was quoted as saying "every person must work and those not working be treated as vagrants." The camps Percy referred to were set up on top of the levees and the scene of a well-orchestrated Red Cross effort to relieve flood refugees. With populations of up to 10,000 refugees, the levee camps became a demonstration of Vice President Herbert Hoover's administrative and organizational expertise. For African Americans, however, the levees were once again drafted into service as areas of refuge much like the Exodusters' reliance on them while waiting passage to Kansas. At one point in April 1927, the local Greenville, Mississippi newspaper reported 6,000 people in the Red Cross camp located on the surrounding levee—most were African American—whereas 4,000 were staying in town. But soon the levees would represent not only refuge from floodwaters but they would also become symbols of the differing treatment for white and African American flood survivors, particularly in Greenville. In other historical accounts highlighting the differential treatment received by white refugees, even the type of transportation accorded to each was different. For example, white refugees were transported downriver to a relief camp in Baton Rouge by covered cabins in steamers, followed by African American refugees in flatboats and barges without the benefit of overhead covering. The differences became more pronounced once they arrived at the relief camp. For the white refugees, they were sheltered in "tree-shaded buildings, the armory or former barracks." For African Americans displaced by the flood—women and children were treated the same as men—they were "housed in small tents in an open field where the mud was ankle-deep-when it rained." Yet, when investigators traveled South to ascertain the validity of stories circulating about the mistreatment of African American flood survivors, the camps in Baton Rouge were "a paradise in comparison" to those in Greenville (Figure 4.2).[8]

Despite increasing criticism from national newspapers, such as the *Chicago Defender,* about the camp conditions for African Americans, Hoover continued to receive kudos for his relief leadership. Many contemporaries at the time credited his leadership—combined with an increasingly professional Red Cross relief effort—as being one of his more noteworthy achievements, leading to his election as president in 1928. His work was touted by many as one editorial stated, "Nation turns to Hoover in time of national disaster" and went on to describe how his background with World War I relief efforts revealed a "mosaic of relief work."

FIGURE 4.2 Barge Loaded with Poor African American Refugees.
Credit: Courtesy of Everett Collection.

For African Americans, however, these flattering portraits could not be further from the truth. Instead, they realized that the indifference shown by Hoover and the Red Cross to the plight of African Americans allowed for the camp brutality and deprivation that many suffered. The camps, policed by the National Guard, reinforced peonage in the Delta as troops prevented African Americans from leaving the levees, assuaging the concerns of Southern planters worried about losing their labor force. It would take later accounts of the 1927 Flood to recognize the racist actions of Hoover and Percy. Both men treated African American refugees grossly different than their white counterparts. Hoover, who would be seeking the presidency, did not want to antagonize Southern voters and Percy, part of an elite Southern planter aristocracy, did not want to lose a labor force.[9]

In their efforts to stop Northern agents from recruiting African American labor, Southern whites employed extreme measures, in an already endangered environment. Reports circulated that African Americans were being held—in some instances by gunpoint—on the levees until the floodwaters receded. According to Greenville resident Jesse Lee McBride when detailing life on the levee, called by one resident the Protection Levee, she said "It was hazardous, with the stock—mules, horses, cows, dogs and everything running around on the levee." She went on to discuss the National Guard presence in Greenville: "They built a station at every corner just about, especially from the levee to the main post office down Washington Avenue. If you came off of the levee and got to the post office and attempted to go back to the levee, you could not go back without a pass. All times you did not know that you had to have this pass from the levee, and the result was when you encountered the officer, or whoever he was, he would also say, "Well, you can't go back." In still another account, Willie Gardner recalled that "black boys and black men were forced to work on the levee with no pay" whereas the whites who worked on the levee were "riders, carrying pistols." In one news account, African American refugees were forced to wear tags citing their names and the plantations where they worked to insure their return and payment for Red Cross rations. Several accounts reveal Southern whites charging displaced African Americans for Red Cross supplies whereas the Red Cross looked the other way although in one of the official advisory reports, this was said to be "contrary to the announced and printed policy of the Red Cross." When reporters or local leaders tried to publicize the abuses and exploitation of African American labor, they were chastised. Still complaints persisted and under mounting pressure, Hoover finally appointed the Colored Advisory Commission to investigate the charges of violence and overall camp conditions. Chairing the committee was Robert Moton of Tuskegee Institute, who reported that levee refugee camp conditions included the presence of armed white guardsmen, which the committee recommended removing. Other recommendations included greater participation by African American refugees in camp oversight and a more equitable distribution of clothing. In response, Hoover convinced the author of the report to soften the harshest indictments of camp life—one claim was that Hoover succeeded in persuading Moton by promising greater African

American political participation after he assumed the presidency—with a final report that still gave Hoover credit for heroic leadership during the crisis.[10]

Supportive of the National Red Cross' action during the flood was an oft-cited article that appeared in *The Nation* in June 1927. Written by the National Association for the Advancement of Colored People (NAACP) assistant executive secretary, Walter White, the article credits the Red Cross with satisfactory relief efforts but is critical of the planters and their treatment of African American laborers. (The Red Cross, however, was not without critics as criticism implied that Red Cross officials were aware of planters selling the relief goods to African Americans.) White cited a *Vicksburg Evening Post* article discussing the insistence of planters that no release of any "family or persons from the camp except with written consent of the landlord from whose plantation the laborers came or on the personal request or authorization of the landlord." To White this continuation of peonage "would recreate and crystallize a new slavery almost as miserable as the old."[11]

Thus, while newspapers and journals documented and editorialized, two flood narratives were evolving. In one narrative, the flood was commemorated as a defining moment in American history. For example, in the late 1930s, Henry Luce, editor of *Life* magazine, commissioned the American artist John Stuart Curry to paint the 1927 disaster as part of Luce's "dramatic scenes in 20th Century American history." In Luce's view, the 1927 Flood was one of those moments that "shaped the national character." In keeping with Luce's perceptions, the portrait that Curry produced shows "a wise and helpful" Hoover at the helm. In an idealized pose, Hoover is at one of the camps overlooking the multitude of refugees, including African American women, men, and children, Red Cross officials, white residents, and farm animals. The flood narrative embodied in the painting is one without Jim Crow or African Americans wearing identification tags. Instead, not only is Hoover depicted as a calm leader in times of crisis but also those affected are seen as a community of dispossessed. Other historical records such as a self-congratulatory Red Cross report include letters of appreciation from flood victims, who embellish the imagery of community that arose during this time of crisis. For example, in one letter of thanks the author writes, "I know of no other more vivid index to the growing sense of brotherhood in mankind than the work you are enabled to do through the consciousness of need and the will to support your work by the public." In contrast, when a Memphis news reporter visited a refugee camp one Sunday, he compared the hymns sung by the white refugees and African Americans. The whites were singing "Jesus Lover of My Soul," asking for protection from the heavens, whereas the African Americans were singing "Swing Low, Sweet Chariot," recounting their passage home to the heavens. The one speaks of hope, whereas the other speaks of resignation. (The resignation, however, that the reporter was referencing might have been more a product of a persistent disproportionate amount of damage that African Americans experienced during times of flooding.) Regardless of the individual narrative, missing from the white flood accounts is the brutality of Southern planters eager to retain their

FIGURE 4.3 River Baptism: Crowd across River Watches Baptism.
Credit: Courtesy of Cleveland Public Library, Special Collections.

labor force or the despair and destitution that so many African American observers encountered when visiting the camps. Again, the intersection of the river with the lives of African Americans and whites prompted accounts and memories that were sharply different (Figure 4.3).[12]

After the flood, in addition to a flood hagiography surrounding leaders such as Hoover, the historical accounts of white Delta residents were often apologies for the treatment received by the African American community or the actions of local leaders such as Leroy and William Alexander Percy. On another level, the flood triggered requests for federal assistance for flood control, a new departure for the traditionally isolated South that began in the 1880s. By November 1927, area representatives from the Mississippi River Flood Control Association in Memphis announced a trip to Washington, D.C. to secure federal support. One of the representatives claimed, "We are about to go on the firing line for the most important campaign in the history of the Mississippi Valley." In a promotional piece by the Mississippi River Flood Control Association, emphasizing the importance of the river and demonstrating the need for federal support, the editor responds to an earlier question as to whether the Mississippi River was a national asset or a liability. Questions such as this reflected a growing trend to frame the river as a national resource and thus a national responsibility. Pleas for federal funding were also found in the Memphis newspaper, which cited all the major Mississippi River floods going back to 1882. After citing the major floods, the article concluded, "The problem of the Mississippi River is a great problem; it is essentially an engineering problem; but it can be solved." The author, who was also a congressional leader, claimed that the river "should be the commercial highway of the nation"—a further justification for federal support. Still other post-flood activities focused on labor and retaining the African American sharecroppers and tenants. In requesting congressional support to rebuild the Delta, one well-known Delta political leader, Walter Sillers, Jr. told his Senator in June 1927, "delta lands without labor are as useless as an automobile without an engine." Not only did the collective memory of the flood differ between African Americans and the white community but also the post-flood narrative diverged as labor and federal flood control relief preoccupied the white community.[13]

In sharp contrast to the narratives by Hoover, professional engineers, Sillers, and other local elites were the remembrances of the African American community in lives that continued to be shaped by the Mississippi. A resilience and at times, empowerment was often at the core of their interactions with the river. For example, while all long-time Delta residents discussed similar initial reactions to news that the levee was going to break such as placing everything they valued in scaffolding in their homes or climbing to the rooftop to escape, some were more creative in their retelling of the 1927 Flood, revealing resilient lives where floods were an expected occurrence. For example, one African American flood survivor, revealing a long history with flooding, recounted how he cut holes in each floor of his home that was perched on blocks. As a result, when he came home, "the house was on the blocks just like I left it, and the floor wasn't buckled, because the water had a chance to come through without any trouble." Others, such as Jesse Lee McBride, challenged authorities when told she could not return to the levee. In her words, she "jumped down in water, which was about waist deep, and waded back down to the levee." And when white Southerners sought to keep African Americans in the South, prominent African Americans such as W.E.B. Du Bois urged African Americans to leave the South immediately:

> We hope that every Negro that can escape from the slave camps guarded by the National Red Cross for the benefit of the big planters of Mississippi and Louisiana and the lynchers of Arkansas will leave this land of deviltry at the first opportunity. Let them ride, run and crawl out of this hell. There is no hope for the black man there.

Du Bois went on to estimate that 75% of the refugees were African American but only received 25% of the relief funds. He also referenced the "refugees from Mississippi Bottoms" and said it was "better to starve in Memphis and Chicago than be slaves in Arkansas and Mississippi." In the narrative presented by Du Bois, the flood was a transformative moment, finally empowering African Americans to leave the South.[14]

Further reflecting the sharp divide between white and African American flood histories—articulated in oral testimonies—was the cultural production of song, prose, and poetry oftentimes relating the 1927 Flood to earlier experiences with the river. One of the most well-known music genres to emerge at the turn of the century was the blues. A unique African American music style, distinct from jazz, the blues drew upon earlier work songs, the "holler," and levee songs or in the words of one blues' musician, "The blues came from behind the mule." Or in the words of another blues musician, W.C. Handy, "The genius doesn't find his music in books and notes but in nature. He hears the music of the storm, the brook, the ocean and makes notes of them."[15] By the 1920s, the popularity of the genre was insured as record companies produced the works (known as race records) of blues artists such such as Bessie Smith, Charley Patton, and Blind Lemon Jefferson, to name a few. Speaking to an African American audience, the blues chronicled the despair and emotion that many felt during the flood. Not everyone agrees about the audience for these records, but their popularity cannot be disputed as by 1926, three of the largest companies were selling five to six million records annually when the entire African American population at that time stood at 15 million. Within the blues industry, flood blues were a separate category with an estimated 40 songs about floods.

One of the most popular flood blues for the 1927 Flood was one by Bessie Smith, known as Empress of the Blues, who experienced the flooding when she was touring in 1927. Blues scholars contend that she probably experienced flooding on one of the Mississippi River tributaries before the major flood. Music scholars believe that her famous "Backwater Blues" reflected that experience in which she captured the unrelenting rain that led to the flood and the loss of homes and displacement for so many African Americans. Her title was also telling because of the common practice of causing a small breach in the levee to relieve some of the pressure in which only the backwaters are flooded. Discussed earlier, the backwaters were often where African Americas had their land.[16] Listed below are the lyrics that not only signaled an individual response to the flood's impact but also would come, like Charley Patton's "High Water Everywhere," to represent the emotions of a community.

"Backwater Blues"

When it rains five days and the skies turn dark as night
When it rains five days and the skies turn dark as night
Then trouble's takin' place in the lowlands at night
I woke up this mornin', can't even get out of my door I woke up this mornin', can't even get out of my door
There's been enough trouble to make a poor girl wonder where she want to go
Then they rowed a little boat about five miles 'cross the pond
Then they rowed a little boat about five miles 'cross the pond

I packed all my clothes, throwed them in and they rowed me along
When it thunders and lightnin' and when the wind begins to blow
When it thunders and lightnin' and the wind begins to blow
There's thousands of people ain't got no place to go.[17]

But Bessie Smith went even further in her lyrical descriptions of the displacement and loss that the flood wrought. In the following song, she gives the river agency as she personalizes what the river and the subsequent flooding meant for her. Not only is she left homeless, but she also states her intention to leave the area and "never come back no more." There is a resilience in the statement, perhaps referencing the migration north that resulted for many flood survivors:

"Homeless Blues"

Mississippi River, what a fix you left me in
Lord, Mississippi River, what a fix you left me in
Mudholes of water clear up to my chin
House without a steeple, didn't even have a door
Plain old two-room shanty, but it was my home sweet home
Ma and Pa got drownded; Mississippi, you to blame
My Ma and Pa got drownded, Mississippi, you to blame
Mississippi river, I can't stand to hear your name
Homeless, yes, I'm homeless, might as well be dead.
Ah, you know I'm homeless, homeless, yes, might as well be dead.
Hungry and disgusted, no place to lay my head.
Wished I was an eagle, but I'm a plain old black crow
I'm gonna flap my wings and leave here and never come back no more.[18]

Adding to Smith's reactions and cited at the beginning of this chapter was Charley Patton's song, "High Water Blues," discussed earlier in the chapter in which he referred to the restrictions imposed on African Americans and the prohibition of leaving the levee camps. But perhaps even more revealing is the song by Kansas Joe and Memphis Minnie where work on the levee by an African American man and the levees' failing were connected:

"When the Levee Breaks"

If it keeps on raining, levee's going to break.
And the water gon' come, and have no place to stay.
Well, all last night I sot on the levee and moaned.
Thinking 'bout my baby and my happy home.
If it keeps on raining, levee's going to break.
And all these people have no place to stay.
Now, look here, mama, what am I to do?

Now, look here, mama, now what should I do?
I ain't got nobody to tell my troubles to.
I worked on the levee, mama, both night and day.
I ain't got nobody to keep the water away.
Oh, crying won't help you, praying won't do no good.
Now crying won't help you, praying won't do no good.
When the levee breaks, mama, you got to move.[19]

Work always punctuated and defined the African American experience with the Mississippi River, as illustrated by earlier cited steamboat and levee songs (Figure 4.4). Together, these songs conveyed different experiences as the river intersected with their lives—in this instance, through work and natural disaster. Through the blues, the flood offered another narrative of suffering and displacement that informed identity, contributing to a long, collective history of the Mississippi River. For African Americans, as seen through the flood blues, the river persisted as a depository of memories that at times inspired but more often saddened. In the midst, however, of these dualistic responses was an enduring quality of resilience. After the 1927 Flood, many were prompted to move north, to

FIGURE 4.4 African American Refugees in Front of their Temporary Tent Homes during the 1927 Mississippi River Flood.
Credit: Courtesy of Everett Collection.

start over. Again, the river and its environs served as a catalyst for flight as it had been for the maroons, the enslaved people who escaped, and the Exodusters.[20]

Adding to song were the works of writers such as Richard Wright with his classic short story, "The Man Who Saw the Flood." Wright's description of the flood is more explicit as he chronicled the unrelenting poverty associated with life along the Mississippi, only to be exacerbated by floods. In Wright's work, he encapsulated the experiences of the African American community as the 1927 Flood became another historical marker in a history of lives interacting with the river. In some instances, African Americans saw the flood as a symbol—washing the slate clean and affording the opportunity to leave and start anew. For example, in one interpretation of Wright's story, the main character, the sharecropper, Tom, had an epiphany after returning home with his family to a silt-filled shack. For a brief moment, Tom contemplated leaving his leased patch of land and the $800 debt he owed the white planter, Mister Burgess, and start over. This moment was captured in Tom's pause when Burgess told Tom that despite losing everything in the flood, he still owed him the $800 debt; however, in the meantime, Burgess would "stake" Tom for the upcoming season. Tom's response was not immediate, prompting Burgess to ask Tom whether he was coming. Seeing no way out, Tom did follow Burgess and reentered a life of peonage but the pause signified a moment of release and the potential for renewal. The theme of starting over was shared by others, as reported to Hoover's Colored Advisory Commission. But for other flood survivors, and this was reiterated in song, the flood was a punishment for past sins.[21]

Regardless of which interpretation was accepted, the treatment of African Americans during the flood exposed a labor system not much different from the antebellum world. As Southern planters in the Yazoo Delta region were accused of peonage, the harsh world of Jim Crow was revealed to the nation. These realities belied a single flood narrative trumpeting a sense of community and shaping the national character through adversity. Instead, the 1927 Flood affected populations differently based upon race and class. These differences, in turn, were conveyed through cultural references of song, prose, and memoirs. For African Americans, the flood was a continuation of a narrative in which the river was always part of the negotiated world between African Americans and Southern whites. This power dynamic resulted in a duality of experiences. In this instance, embracing the extremes of renewal and despair, exodus and bondage were once again the discourse between African Americans and the river. Realizing these experiences shaped a collective memory still celebrated and lamented. But the aftermath of the flood also predicated subsequent histories as the Mississippi represented dual roles of liberator and oppressor—the former offering a means to escape north, the latter further imprisonment in peonage. Or the river represented renewal and tyranny—the former being a sense of starting anew as the floodwaters receded, the latter a return to silt-filled homes with less than earlier.

This collective memory was further enhanced with more flooding. For African Americans in the 1920s, the Mississippi River continued to oppress as a 1929 account illustrates:

> It is Mississippi. "The River" is on its annual rampage. Wastes of water spread over the landscape. Houses are submerged up to the second story. Boats have replaced automobiles and buggies . . . At any time, any and every Negro may be impressed to save the levees. . . Everywhere there is unrest and an atmosphere of fear and suspense.[22]

Recognizing the Mississippi's "annual rampage" implied an understanding of the river, which makes the 1927 Flood less a singular event for African Americans living by the river. While there is no question, the flooding was worse than previous floods, and for many the same themes endured—the ever-present threat of lost income, displacement, and the countless abuses of a social system governed by Jim Crow. Thus, when assessing the historical significance of the 1927 Flood for African Americans, again a different history emerged. For African Americans, the flood altered their lives, prompting the migration north; leaving lives based on an agricultural, rural economy. Granted that the major movement north occurred after 1940 as 77% of the African American population still lived in the South in 1940 with 49% living in rural areas, black migration north to cities such as Chicago had begun. For a 60-year period, from 1910 to 1970, six and one-half million African Americans moved north. For the period from 1910 to 1940, one and one-half million left the South for Northern opportunities the 1927 Flood was one of the catalysts for the move.[23]

More catastrophic floods did occur, again disproportionately affecting African Americans. In 1937, the Mississippi River Basin experienced a 1,000-year flood, what one scholar labeled the "Ohio Mississippi Disaster of 1937." This flood was caused, in part, by levees that had been constructed so high that the river's overflow had nowhere to spread out and the smaller tributaries were blocked from entering the main channel. In other words, the "Mississippi backed up into its tributaries as if cascading over a waterfall." As in earlier floods, for many African Americans working the land, their farms were located on the river bottoms, which were always the first to be flooded. For those African Americans living in urban areas, they often resided in neighborhoods along the waterfront, which also would be the first to be inundated.[24] Echoing the reactions of blues artists after the 1927 flood, the 1937 flood was also memorialized through the music of blues artist, Big Bill Broonzy. Thus, in "Southern Flood Blues," Broonzy lamented not having anywhere to go as "water was comin' in my door." Other parallels with the 1927 Flood inferred a long history with the river and its recurring floods with several references. First, the flooding and his reaction was historical as he sang, "It was the old high (?) river, tellin' us to get ready and go." Second, another twentieth-century refrain—combining hope with disdain for the South—was Broonzy's promise: If he could get away, he "will never come here again." Finally, Broonzy's song depicted a vivid scene of the devastation and human suffering that all flood victims endured but in disproportionate numbers for those African Americans living on the bottomlands or urban waterfronts.[25]

Thus, by the 1930s, African Americans were framing their history with the Mississippi River through several outlets, succeeding in reaching national audiences. Discussed earlier, blues music was one means of chronicling a history, characterizing major events such as the 1927 Flood and exposing the systemic racism in Southern society. Paralleling the popular culture of the blues that reached millions were the exposés of publications such as *The Crisis* and African American newspapers such as the *Chicago Defender.* An awareness of living conditions for those dwelling in the Mississippi River Basin became heightened on an unprecedented level. Literary works, such as Richard Wright's story discussed earlier, contributed to a new environment for African Americans where the nostalgia of the Old South, the mythology of the Lost Cause, came under scrutiny. Of course, it would be decades before substantial changes were introduced but for African Americans living along the Mississippi River and experiencing frequent flooding, the 1927 Flood invited an exposure that was groundbreaking. Slowly, an African American narrative of the Mississippi River was gaining a national audience.

Further contributing to a national consciousness while also revealing another area where African American lives intersected with the river was the work performed at the levee camps. By the late 1920s and early 1930s, levee building reached new lows for African Americans working in camps that were subcontracted to local contractors by the U.S. Army Corps of Engineers. After the 1927 Flood, federal funding for levee building and maintenance increased and many African Americans sought levee work to supplement their incomes. Before the work of digging and dredging became mechanized, sharecroppers often labored as muleskinners as they were known for their skills working with the mules. Similar to working conditions in the levee camps operated through the convict lease system, the post-1927 camps became well-known theaters of violence and corruption. Since the federal government allowed local contractors to hire and supervise laborers working on the levees, without any oversight, abuses were rampant. Hearing of the appalling camp conditions in late 1932, the NAACP sent staff member, Helen Boardman, to investigate. In her report, she wrote "all along the Mississippi . . . Negro workers [were] crowded into floorless tents." She also stated that laborers worked 12-hour days, seven days a week for an average of ten cents an hour. Another complaint was the practice of the commissaries charging exorbitant prices and making fortunes from the workers' earnings. The NAACP report was sent to President Hoover along with other administration officials and senators. Unfortunately, no action was taken. The NAACP, however, did not end their investigations. Funded by the NAACP, the well-known journalist, George S. Schuyler, described what he found in two levee camps along the river. The letter written to Walter White, then secretary of the NAACP, reported on conditions at a camp near Waterproof, Louisiana and another by Locust Ridge. From conversations with the men, Schuyler learned that the wages ran from $1.50 to $1.75 a day for dragline work and $2.50 for work laying mattresses. The workday was never less than ten hours and sometimes the men were expected to work into the night with no overtime pay, only an hourly wage. In his conclusions to White, Schuyler said the main complaint was the low wages and the fact that room and board was not provided. Instead, the men were expected to pay for room and board out of the meager wages. An even more damming account of levee work appeared in April 1933 in *The Crisis.* Titled "Mississippi Slavery in 1933," by civil rights activist Roy Wilkins, the article critiqued a levee camp system where the workers had little cash of their own, always working within a credit system where they found themselves in constant debt. The camps that Wilkins visited, again through NAACP support, were under the aegis of the U.S. Corps of Engineers in which Wilkins censured with the following: "It is no exaggeration to state that the conditions under which Negroes work within federally-financed Mississippi levee construction camps approximate virtual slavery." His report echoed the complaints that Schuyler heard from the men in which they bemoaned the long hours, seven-day work weeks, and irregular pay. More responsive this time, the Senate in 1933 did "set minimum standards of working conditions and wages for all unskilled labor engaged in flood control." Abuses, however, persisted and the NAACP acted as a watchdog throughout the 1930s. When conditions worsened, the organization alerted the War Department and in some instances, action was taken. By the 1940s, however, most levee construction had been mechanized with drag lines replacing the mule teams, ending a centuries-old labor-driven intersection between African Americans and the Mississippi River.[26]

Despite the arrival of machinery to build and maintain the levees, blues artists kept the memories of the levee camps alive for the African American community. The levees became part of the infrastructural memory of the river as blues musician, Son House, illustrated in his well-known "Levee Camp Moan." In this composition, Son House sang about a "good looking woman" who was only interested in his paycheck—money earned from his levee job. The song described her as always waiting on the boat landing on payday. But when he came home without a paycheck, in his words, "She told me she couldn't use me no more." With the Son House recording, the levee is a backdrop, a part of everyday life and culture in the Mississippi Delta, indicating how integral levees and levee work was to those living along the river.[27] But another part of levee camp life—exploitation by the contractor—was enshrined through Gene Campbell's blues contribution:

"Levee Camp Man Blues"

These contractors, they are getting so slack
These contractors, they are getting so slack
They'll pay you half of your money and hold the other half back

There ain't but two men that get paid off
There ain't but two men that get paid off
That's the commissary clerk and the walking boss

I see somebody coming down to the water trough
I see somebody coming down to the water trough
I know it ain't the contractor, it's that doggone walking boss

A levee camp mule and a levee camp man
A levee camp mule and a levee camp man
They work side by side, and it sure is man for man

A levee amp man ain't got but two legs you know
A levee camp man ain't got but two legs you know
But he puts in the same hours that a mule do on four

I wouldn't drive no four-mule team
I wouldn't drive no four-mule team
For no doggone contractor I ever seen

Men on the levee hollering "Whoa Haw Gee"
Men on the levee hollering "Whoa Haw Gee"
And the women on the levee camp hollering "Who wants me?"[28]

In Campbell's song, he captured not only the unfair system of wages and never-ending debt but also the brutalizing treatment of labor: Man and mule were treated the same.

Finally, through Lonnie Johnson's song, working on the levees would always be associated with the 1927 Flood. In the following lyrics, Johnson lamented the choices of either working on the levee or going to jail:

"Broken Levee Blues"

I wants to go back to Helena, the high waters got me bogged.
I wants to go back to Helena, the high waters got me bogged.
I woke up early this mornin', a water hole in my back yard.

They want me to work on the levee, I have to leave my home.
They want to work on the levee, that I have to leave my home.
I was so scared the levee might break out and I may drown.

The water was round my windows and backin' all up in my door.
The water was all up 'round my windows and backin' all up in my door.
I'd rather to leave my home 'cause I can't live there no more.

The police run me all from Cairo, all through Arkansas.
The police run me all from Cairo, all through Arkansas.
And put me in jail, behind those cold iron bars.

> The police say work, fight, or go to jail, I say I ain't totin' no sack.
> Police say work, fight, or go to jail, I say I ain't totin' no sack.
> And I ain't buildin' no levee, the planks is on the ground and I ain't drivin' no nails.[29]

But violence was not confined to the plight of the levee worker or flood survivors as lynchings were still a reality in the South. According to an investigation undertaken by Walter F. White, who led the NAACP from 1931 to 1955, he reported 41 lynchings and eight race riots over a ten-year period covering the 1920s. While his investigations covered the entire South, of particular interest was a trip to Phillips County, Arkansas, a rich cotton-producing area, part of the Arkansas Delta located along the Mississippi River. White's interest in the area stemmed from an initial report that African Americans in the state planned to "massacre all the white people of the State." Instead, White found that African Americans in the region had created a cooperative as defense against the "exploitation by white landlords, merchants, and bankers, many of whom openly practiced peonage [slavery]." In his inquiries, White learned that similar to the treatment of sharecroppers throughout the Delta, in Phillips County, sharecroppers were also subject to an unfair credit system and never allowed an itemized statement of their charges and earnings. Those that protested these tactics had been lynched. In fact, during White's investigation, he was forced to leave town quickly or else face an angry mob. Despite exposés such as White's, lynchings continued in the South. According to NAACP records, from 1882 to 1968, the South saw 4,743 lynchings and out of that total, the majority occurred in Mississippi with 581 lynchings. The lynchings were reminiscent of earlier acts of violence against those living in the Mississippi River Basin, dating back to the treatment of the maroon, San Malo, and the survivors of the 1811 Revolt.[30]

Returning to the years after the 1927 Flood, not only were there more floods—despite the U.S. Army Corps of Engineers' levee-building efforts—but also with the advent of the Great Depression and Franklin Roosevelt's election to the presidency, New Deal programs provided more relief to the South. Federal support, however, such as farm subsidies ultimately hurt African American sharecroppers. Since the Agricultural Adjustment Act paid farmers subsidies not to plant crops, this often left African American sharecroppers without work. By the 1930s, African Americans engaged in agriculture were still struggling. In one study, the author found that in Mississippi alone, almost three-quarters of the population were tenants. The same study found that "as a whole, rural areas of the South have remained nearer the economic margins of subsistence than any other major region." Despite the grim statistics, media productions such as the federally financed film, "The River," by Paul Lorentz, only spotlighted the plight of white sharecroppers, with scenes illustrating how past flooding has affected them. The film, which received numerous kudos, which was a commentary about the Mississippi River, was also a commemoration of the Tennessee

Valley Authority, and the expertise of the federal government to control nature. While Lorentz evoked an image of unity surrounding New Deal programming, there were no representations of African Americans and the suffering they endured through Mississippi River floods. The film not only was, undoubtedly, in part, a product of New Deal efforts to ensure Southern support for federal programming but also revealed what Carolyn Finney has termed "landscapes of exclusion."[31]

Paralleling a national conversation celebrating the New Deal Program, enthralled by the perceived ability of technology to reconstruct nature and facilitate the modern nation-state, African Americans also experienced successes and failures in their centuries-old kinship with the river. First, the resilience demonstrated by African Americans in their ongoing relationship with the river endured. The choice to move north—whether prompted by the 1927 Flood or later events—was in many ways similar to the choices of the Exodusters, the enslaved who sought freedom through escape to the north, or the maroons who found refuge in the river's swamplands. In all these examples, the riverine landscape incorporated flight in the collective memory. During the 1927 Flood, however, the river acted as both punisher and liberator, much like the role it played during the days of slavery. Still another continuity was the decision by many to start over as Richard Wright's main character does. This resilience, which would also be witnessed by those who were adamant about returning to New Orleans after Katrina, characterized a unique relationship with the river and its surroundings. To the white planters and politicians who framed their response to the flood as an opportunity to engineer the river into a 9-foot channel, free from flooding while serving commerce, a historical drama filled with an expanding confidence in technology's potential, African Americans living on the low-lying bottomlands had a different perspective. The river and the latest flood represented either a highway north to a better life or an opportunity to start over. But underlying both choices was the recognition of the constancy of the river—flooding and the attending devastation were part of the Mississippi's genetic code. Thus, while the 1927 Flood was disastrous, causing more suffering and loss than earlier floods, for African Americans living along the Mississippi, disproportionate suffering had been the norm. Instead, the more significant changes were the national audience that the flood generated and the cultural production of music and prose that attracted the attention of millions. The 1927 Flood expanded and enshrined a collective memory, informing an identity that rested upon the histories of millions of African Americans who had lived along the river since the early 1700s.

But in returning to the 1920s, the river remained central to the African American community, lending an immediacy that contrasted with the post-flood discourse for Southern whites who associated floods with the need for more control of the river through federal support. Multiple narratives persist up to the present as the flooding associated with Hurricane Katrina affected an impoverished African American population to a much greater degree than the white residents in the vicinity of New Orleans. The historical memory of the river continues as efforts to rebuild New Orleans reveal a cultural baggage tied to the river. Although written in 1922, Langston Hughes' poem, "The Negro Speaks of Rivers" still resonates today as the dynamics between rivers and cultures are part of a trajectory that informs identity, based on race and class, often reflecting the fissures of the larger society.

"The Negro Speaks of Rivers"

I've known rivers:
I've known rivers ancient as the world and older than the flow
Of human blood in human veins

My soul has grown deep like the rivers
I bathed in the Euphrates when dawns were young
I built my hut near the Congo and it lulled me to sleep.
I looked upon the Nile and raised the pyramids above it.
I heard the singing of the Mississippi when Abe Lincoln went
down to New Orleans, and I've seen its muddy bosom turn
all golden in the sunset.
I've known rivers:
Ancient, dusky rivers.
My soul has grown deep like the rivers.

Notes

1 Charley Patton (1929) "High Water Everywhere," Part 1 may be accessed at https://genius.com/Charley-patton-high-water-everywhere-part-1-lyrics.
2 According to Francis Davis, Patton's "High Water Blues" was the most "vivid" of blues songs about the 1927 Flood with lyrics that were "observational and autobiographical." See Davis, *The History of the Blues: The Roots, the Music, the People from Charley Patton to Robert Cray Francis Davis* (New York: Hyperion, 1995):97–98. In looking at the communal aspect of blues music, in Cynthia Gooch's dissertation, she devoted a chapter to examining the blues, in which she convincingly argued that "representations of the Mississippi River in blues music express both individual and communal hardship during specific historical moments." The 1927 Flood would be one of those historical moments. See Gooch, ""I've Known Rivers:" Representations of the Mississippi River in African American Literature and Culture" Ph.D. Dissertation (University of Kentucky, 2019):89 accessed at https://uknowledge.uky.edu/cgi/viewcontent.cgi?article=1125&context=english_etds. An earlier discussion of the blues and how they remained communal property, vehicles for individual and group expression, can be found in the classic by Lawrence W. Levine, *Black Culture and Black Consciousness* (1977; Oxford: Oxford University Press, 2007):217–239.
3 Barbecue Bob (1927) "Heavy Water Blues," may be accessed at https://genius.com/Barbecue-bob-mississippi-heavy-water-blues-lyrics
4 For general histories of the 1927, see John M. Barry, *Rising Tide: The Great Mississippi Flood of 1927 and How It Changed America* (New York: Simon & Schuster, 1997); Pete Daniel, *Deep' N As It Come: The 1927 Mississippi River Flood* (New York: Oxford University Press, 1977; Patrick O'Daniel, *When the Levee Breaks: Memphis and the Mississippi Valley Flood of 1927* (Charleston: The History Press, 2013); David Cohn,

Where I Was Born and Raised (Notre Dame: University of Notre Dame Press, 1935). Environmental historian Christopher Morris contends, "The 1927 flood was, and by some measures remains, the nation's greatest natural disaster, remembered in photos, songs and film." See Morris, *The Big Muddy: An Environmental History of the Mississippi and Its Peoples from Hernando De Soto to Hurricane Katrina* (New York: Oxford University Press, 2012):165. For an excellent overview of the year 1927 focusing on media representations, see Susan Scott Parrish, *The Flood Year 1927: A Cultural History* (Princeton: Princeton University Press, 2017). Two works that have focused on the 1927 Flood through blues music are Richard M. Mizelle, Jr., *Backwater Blues: The Mississippi Flood of 1927 in the African American Imagination* (Minneapolis: University of Minnesota Press, 2014) and Stephanie Deneve, "Representing Environmental Emergency as Social Emergency: The Great Mississippi Flood of 1927 in Blues Songs from Louisiana and Mississippi" *Catalogue Des 559 Revues* (2 August 2021) accessed at https://journals.openedition.org/erea/11658

5 Mary Hamilton, *Trials of the Earth: The Autobiography of Mary Hamilton* (Oxford: University of Mississippi Press, 1992):52; Mikko Saikku, *This Delta, This Land: An Environmental History of the Yazoo-Mississippi Floodplain* (Athens: The University of Georgia Press, 2005). Another deterrent to earlier settlement in the Delta region was disease as the South Carolinian planter James Ruffin observed in 1833, when he described the Yazoo country as being "very sickly, though it is very productive" and then went on to write, "the negroes die off every few years, though it is said that in that time each hand also makes enough to buy two more in his place." See James H. Ruffin to Thomas Ruffin, 30 April 1833, *The Papers of Thomas Ruffin*, Vol. II (Raleigh: Edwards & Broughton Printing Co., 1918):77.

6 See Chapter 3 for a discussion of the founding work of the Mississippi River Commission. Daniel, *Deep' N As It Come: The 1927 Mississippi River Flood*, p. 5; *Commercial Appeal* (Memphis), 8 May 1927. In reviewing the 11 worst floods that occurred in the Mississippi River Basin from 1858 to 1927, the 1927 flood was determined the most damaging. See James C. Cobb, *The Most Southern Place on Earth: The Mississippi Delta and the Roots of Regional Identity* (New York: Oxford University Press, 1992):129.

7 *Daily Democrat-Times* (Greenville), 21 April 1927; Winchester Davis Interview, p. 11, Oral History Collection, William Alexander Percy Library, Greenville, Mississippi; Daisy Miller Greene Interview, p. 13, Oral History Collection, Percy Library; Barry Smith Interview, Federal Writers' Project: Slave Narratives, 1936–1938, Mississippi (Bedford: Applewood Books, rpt.):132–133.

8 *Daily Democrat-Times* (Greenville), 26 April 1927; *Daily Democrat-Times,* 16 May 1927; *Daily Democrat-Times*, 26 April 1927; Langston Hughes, *The Collected Works of Langston Hughes, Fight for Freedom and Other Writings on Civil Rights*, Vol. 10, ed. Christopher C. DeSantis (Columbia: University of Missouri Press, 2001):88–89. In Hughes' recounting of the 1927 Flood, he refers to subsequent investigations in the 1930s in the section, "Old Man River."

9 *Commercial Appeal* (Memphis), 9 May 1927. For an excellent discussion regarding Hoover's political ambitions and how his work overseeing 1927 Flood relief was influenced by his drive to become president, and thus not alienate Southern Democrats, see Myles McMurchy, ""The Red Cross is Not All Right!": Herbert Hoover's Concentration Cover-Up in the 1927 Mississippi Flood," *Yale Historical Review* 5 (Fall 2015):87–114. For a discussion of peonage in the Delta, see Pete Daniel, *The Shadow of Slavery: Peonage in the South, 1901–1969* (Urbana: University of Illinois Press, 1972); Richard McKinley Mizelle, Jr. "Backwater Blues: The 1927 Flood Disaster, Race, and the Remaking of Regional Identity, 1900–1930" Ph.D. diss., Graduate School-New Brunswick Rutgers, State University of New Jersey, 2006; Robyn Spencer, "Contested Terrain: The Mississippi Flood of 1927 and the Struggle to Control Black Labor" *The Journal of Negro History*, Vol. 79, No. 2 (Spring, 1994):170–181.

10 White Southerners always worried about African American migration north and the impact on their labor forces. See *The Crisis*, Vol. 16, No. 3 (June 1918):71; Hortense Powdermaker, *After Freedom: A Cultural Study in the Deep South* (New York: Viking Penguin, 1939; rpt. 1993). *Chicago Defender*, 7 May 1927; *Daily-Democrat Times*, 14 May 1927; *Chicago Defender*, 4 June 1927; Jesse Lee McBride Interview, 13, Willie D. Gardner Interview, p. 18, Oral History Collection, Percy Library. American National Red Cross, *The Final Report of the Colored Advisory Commission Appointed to Cooperate with the American National Red Cross and the President's Committee on Relief Work in the Mississippi Valley Flood Disaster of 1927* (Washington: American National Red Cross, 1929). Throughout his article, McMurchy is critical of both the Red Cross and Hoover for their role in the treatment of African Americans in the relief camps. See McMurchy, ""The Red Cross is Not All Right!": Herbert Hoover's Concentration Camp Cover-Up in the 1927 Mississippi Flood." For further discussion of the abuses uncovered at the relief camps, see Hughes, *Collected Works*, Vol. 10, pp. 88–89.

11 Walter White, "The Negro and the Flood" *The Nation*, Vol. 124, No. 3233 (22 June 1927):688–689. In Gooch's work, she discusses how the 1927 Flood "allowed such exploitation (black labor) to continue and increase." Gooch, "I've Known Rivers," p. 76.

12 Charles C. Eldredge, *John Stuart Curry's Hoover and the Flood: Painting Modern History* (Chapel Hill: University of North Carolina Press, 2007):7–8, 65; *The Mississippi Valley Flood Disaster of 1927: Official Report of the Relief Operations* (Washington: American National Red Cross, 1928):148. In David Evans' words, the difference in the two hymns offers "no clearer expression of the difference in attitudes toward the flood." David Evans, "High Water Everywhere: Blues and Gospel Commentary on the 1927 Mississippi River Flood," in *Nobody Knows Where the Blues Come From*, ed. Robert Springer (Jackson: University Press of Mississippi, 2006):10. In DENÉVE's article, in examining the African American response to the 1927 Flood, she emphasized their representations of the flood through blues music. She juxtaposed this response with that of local elites who declared war on the river. While accepting DENÉVE's thesis regarding how African Americans understood and represented the flood, my text differs in that the flood is not viewed as a singular event but part of a long history between African Americans and the river. The blues, derived from earlier African American popular culture outlets such as the songs of the roustabouts, represent a long-standing means of interpreting and conveying that history. See DENÉVE, "Representing Environmental Emergency as Social Emergency: The Great Mississippi Flood of 1927 in Blues Songs from Louisiana and Mississippi."

13 David L. Cohn, *Where I Was Born and Raised* (Notre Dame: University of Notre Dame Press, 1935); William H. Haas, "The Mississippi River—Asset or Liability" *Economic Geography*, Vol. 7, No. 3 (July 1931):252–262; *The Flood of 1927: Mississippi River and Its Tributaries* (Mississippi River Flood Control Association, 1927); Walter Sillers Papers, Bulletin 129, Delta State University Library, Cleveland, Mississippi; *Commercial Appeal*, 8 May 1927; Walter Sillers, Jr. to Senator Pat Harrington, 9 June 1927, Walter Sillers Jr. Papers. In Cobb's work, *The Most Southern Place on Earth: The Mississippi Delta and the Roots of Regional Identity*, he observed that the 1927 Flood and the subsequent requests for federal support prepared the South for the Great Depression and New Deal programs. See Cobb, p. ix.

14 Will Moore Interview, p. 29, Emery John Gipson, Sr. Interview, p. 8, Jesse Lee McBride Interview, Oral History Collection, Percy Library W.E.B. Du Bois, *The Crisis* (July 1927):168.

15 Giles Oakley, *The Devil's Music: A History of the Blues* (London: DeCapo Press, 1997):10; W.C. Handy, *Father of the Blues: An Autobiography* (1941; Boston: Da Capo Press, 1969):ix.

16 For the discussion of the blues, their origin and influence, I owe a large intellectual debt to David Evans for his earlier work on flood blues generated by the 1927

Mississippi River Flood. See Evans, "High Water Everywhere: Blues and Gospel Commentary on the 1927 Mississippi River Flood" *Nobody Knows Where the Blues Come From: Lyrics and History,* ed. Robert Springer (Jackson: University Press of Mississippi, 2006); Evans, *Big Road Blues: Tradition and Creativity in the Folk Blues* (Berkeley: University of California, 1982). Paul Oliver's pioneering work also supports my thesis, as he contended that the blues offered insights into a shared history of African Americans eking out a living on the backwaters of the volatile, unpredictable, and ultimately unsubdued Mississippi River but with the caveat that not all Black Americans identified with the blues; however, countering his argument is the reality that the large sales of blues records in the 1920s testified to a substantial following. See Paul Oliver, *Blues Fell This Morning: Meaning in the Blues* (Cambridge: Cambridge University Press, 1960). The scholarship of Angela Y. Davis and her thesis that the blues music of female musicians such as Bessie Smith was a form of social protest and thus revealed insights into everyday African American life also supports this chapter. In her words, "blues preserved and reflected a cultural consciousness." See Angela Y. Davis, *Blues Legacy and Black Feminism: Gertrude "Ma" Rainey, Bessie Smith, and Billie Holiday* (New York: Vintage, 1999).

17 Bessie Smith (1927) "Backwater Blues" may be accessed at https://genius.com/Bessie-smith-backwater-blues-lyrics

18 Bessie Smith (1927) "Homeless Blues" may be accessed at https://www.google.com/search?q=lyrics+to+bessie+smith%2C+homeless+blues&rlz=1C1SQJL_enUS783US783&oq=lyrics+to+bessie+smith%2C+homeless+blues&aqs=chrome.69i57j0i22i30.7435j0j15&sourceid=chrome&ie=UTF-8

19 Memphis Joe and Kansas Minnie (1929) "When the Levee Breaks" may be accessed at https://genius.com/Kansas-joe-mccoy-and-memphis-minnie-when-the-levee-breaks-lyrics

20 In assessing the role of the river in the lives of African Americans, Cynthia Gooch observed, "It does not simply hold oppression and pain in its waters; it holds nostalgic oral histories, newfound meanings, and empowering reconstructions." See Gooch, "I've Known Rivers," p. 175.

21 Richard Wright in "The Man Who Saw the Flood" *Eight Men* (New York: Thunder's Mouth Press, 1940):110–117. For an excellent discussion of this short story, see William Howard, "Richard Wright's Flood Stories and the Great Mississippi River Flood of 1927: Social and Historical Backgrounds" *The Southern Literary Journal*, Vol. 16, No. 2 (Spring 1984):44–62. David Evans, "High Water Everywhere: Blues and Gospel Commentary on the 1927 Mississippi Flood" *Nobody Knows Where the Blues Come from*, p. 10.

22 Robert W. Bagnall, "The Present South" *The Crisis*, Vol. 36, No. 9 (September 1929):303.

23 Nicholas Lemann, *The Promised Land: The Great Black Migration and How It Changed America* (New York: Vintage, 1991):6. Another development that spurred migration was the increasing mechanization of cotton farming.

24 David Welky, *The Thousand-Year Flood: The Ohio Mississippi Disaster of 1937* (Chicago: University of Chicago Press, 2011):6, 10, 193. Welky stated that while there were some instances of mistreatment in the 1937 Flood, the administration of Franklin Roosevelt was an improvement over that of Calvin Coolidge. See Welky, *The Thousand-Year Flood,* p. 193.

25 Big Bill Broonzy (1937) "Southern Flood Blues" accessed at https://genius.com/Big-bill-broonzy-southern-flood-blues-lyrics

26 Michael McCoyer, "Levee Camps" *Mississippi Encylopedia* (July 2017) may be accessed at https://mississippiencyclopedia.org/entries/levee-camps/; George S. Schuyler to Walter White, 23 December 1932, NAACP Records, Manuscript Division, Library of Congress, may be accessed at https://www.loc.gov/exhibits/naacp/the-great-depression.html; Roy Wilkins, "Mississippi Slavery in 1933" *The Crisis* (April 1933) may be accessed at https://assets.xulastory.com/e027718683d9f1ff859e45e0503216af;

Hughes, *Collected Works,* pp. 90–91. For an excellent discussion of the NAACP investigations into the levee camps, see Richard L. Mizelle, "Black Levee Camp Workers, The NAACP, and the Mississippi Flood Control Project, 1927–1933" *The Journal of African American History*, Vol. 98, No. 4 (Fall 2013):511–530. Levee work and the work songs associated with it also revealed perceptions of African American male identity; see Michael McCoyer, "Rough Mens" in "the Toughest Places I Ever Seen: The Construction and Ramifications of Black Masculine Identity in the Mississippi Delta's Levee Camps, 1900–1935" *International Labor and Working-Class History*, No. 69, Issue 1 (Cambridge University Press, Spring 2006):57–80.

27 Son House (1941) "Levee Camp Moan" accessed at https://genius.com/Son-house-levee-camp-moan-lyrics

28 Gene Campbell (1930) "Levee Camp Man Blues" accessed at https://www.musixmatch.com/lyrics/Gene-Campbell/Levee-Camp-Man-Blues

29 Lonnie Johnson (1928) "Broken Levee Blues" accessed at https://songmeanings.com/songs/view/3530822107858852992/

30 Walter F. White, "I Investigate Lynchings," *American Mercury* (January 1929) accessed at http://nationalhumanitiescenter.org/pds/maai3/segregation/text2/investigate-lynchings.pdf; NAACP, *History of Lynching* in America accessed at https://naacp.org/find-resources/history explained/history-lynching-america

31 Nancy L. Grant, *TVA and Black Americans Planning for the Status Quo* (Philadelphia: Temple University Press, 1990):xxii; Charles S. Johnson, *Growing up in the Black Belt: Negro Youth in the Rural South* (Washington: American Council of Education, 1941):43; Finney cites the work of other scholars when using this term. See Carolyn Finney, *Black Faces, White Spaces: Reimagining the Relationship of African Americans to the Great Outdoors* (Chapel Hill: University of North Carolina Press, 2014):7.

5

MEMORY PERSISTS

Community and Hurricanes

FIGURE 5.1 Fazendeville-Jean Lafitte National Historical Park and Preserve.
Credit: Courtesy of Jean Lafitte National Historical Park and Preserve, Barataria Preserve and Chalmette Battlefield and National Cemetery.

By the mid-1950s, the Lower Mississippi Valley was home to multiple African American communities (Figure 5.1). Fazendeville, also known as "the Village," was one of those communities. Often forgotten by historians, for almost 100 years Fazendeville boasted an African American community of 44 to 50 families. A short distance from the nucleus of the city of New Orleans, the Village was located on the east side of a Mississippi River levee once sandwiched between the

DOI: 10.4324/9781315617077-6

Chalmette National Cemetery and Monument and part of St. Bernard Parish. Because of its isolation, the self-sufficient village was able to escape the race tensions that plagued greater New Orleans. Although out of the-way, the community housed a schoolhouse—that doubled as a dance hall in the evenings—grocery stores, two churches, bars, and used the backfields for a baseball diamond. Once thriving, the residents relied upon the riverine environment and caught crawfish from a nearby canal, grew vegetables, raised chickens, hunted rabbits, and gathered pecans from surrounding pecan trees.[1]

Another community that many African Americans called home was the nearby Ninth Ward, a low-lying area once dominated by cypress swamps. By the mid-1800s, freed African Americans and immigrant workers started moving to the area, attracted by low prices for the low-lying lands bordering cypress swamps. From the onset as New Orleans grew, the higher ground that was less susceptible to flooding would be home to prosperous whites while African Americans and newly arrived immigrants lived on the flood prone lower-lying tracts. Yet, despite the geographical vulnerability to flooding in the Ninth Ward, the area offered a tangible future: the prospect of land ownership. Residents also derived benefits from the riverine environment such as nearby fishing and the opportunity to have gardens where they grew okra and other vegetables. Up until the early 1960s, both Fazendeville and the Ninth Ward were active, self-contained communities where largely African American populations cultivated an environment that fostered a sense of place with communal traditions. All of this dramatically changed, however, by the mid-1960s. By 1963, the Village no longer existed as homes were razed under the auspices of eminent domain and the site was reclaimed to commemorate the 1815 Battle of New Orleans. In response, many Fazendeville residents moved to the Ninth Ward. But by September 1965, the Ninth Ward had also experienced threats to its existence with massive flooding and subsequent destruction, prompted by Hurricane Betsy. Less than 50 years later, the Ninth Ward's existence would again be questioned by the storm surge of another hurricane—Katrina.[2]

Before these communities were threatened by extinction—prompting multiple accounts lamenting the outcomes of hurricanes or the actions of federal officials—another narrative exists. In this coexisting narrative, vibrant, robust communities lying along the Mississippi River inform collective memories, lending not only a sense of place but also a tradition of community support. Beginning with Fazendeville's history, in 1857, Jean-Marie Fazende, a free man of color and a New Orleans merchant, inherited a parcel of land that sat on the site where the famed Battle of New Orleans was fought during the War of 1812. The land had previously been part of a rice plantation and enjoyed a geographical setting—lying next to a natural levy—that rarely flooded. After the Civil War, Fazende sold the land, representing 33 lots, to formerly enslaved African Americans. The community grew throughout the nineteenth century, although never exceeding 50 families. In recollections from former residents and their descendants, they described a family-like environment with a long history as one resident claimed his family lived in Fazendeville for 100 years within a community where everyone knew each other. Adults recounting childhood memories revealed how all the grown-ups always knew what the children were doing, whether within their own family or a neighbor's. Further, the idea of helping each other was not only extended to community members but also anecdotal accounts by outsiders such as that of a delivery man recalled villagers assisting him when his truck broke down on the village road. The mainstay of Fazendeville, however, was the Battle Ground Baptist Church. Life revolved around the church, which would be rebuilt several times. One of the most significant Church events was the ritual of baptism in which congregants walked to the nearby Mississippi River for their immersion. According to one scholar who interviewed numerous Fazendeville residents, "river baptism was one of the most essential spiritual events of the year." In describing the event, she noted that as members walked to the Mississippi River, they sang either "Take Me to the Water," or "Wade to the Water." Baptism remained a prominent memory and long after the Church was moved from Fazendeville to the Ninth Ward, members still recalled the baptismal ceremony. Extant photographs have captured the baptismal ceremony testifying to another linkage with the river, dating back centuries, illustrating its centrality to African Americans living in Fazendeville.[3]

The community of Fazendeville remained a fixture in the Lower Mississippi Valley until the houses were condemned through eminent domain and the entire town was torn down in 1963, almost 100 years after its establishment. The official reason for the community's condemnation was the desire of civic boosters and local historic preservationists, such as the well-known Martha Robinson, combined with the wishes of the LaFitte Park Superintendent, Kyle K. Linch to reconfigure the Fazendeville site into a memorial commemorating the Battle of New Orleans on the eve of its sesquicentennial. Competing narratives coexisted—one that privileged a military victory that cemented the existence of a new republic and another that revealed a self-sufficient African American community that escaped the racial tensions of the late nineteenth and twentieth centuries while sustaining an agricultural way of life and supporting a robust, tight-knit community. Conflicting, however, with the official reason, citing the commemoration of the Battle of New Orleans, were the speculations of many residents and later scholars that the homes were condemned in anticipation of new laws regarding integration and voting rights. Adding to these suspicions and further worsening the situation was the treatment residents received when their homes were seized. When condemning land under the law of public domain, residents are to receive fair compensation for their property, which was not the case for those living in Fazendeville. While other homes in St. Bernard Parish were selling for an average of $16,000, the villagers only received an average of $4,256—hardly enough to start over in a new community. A lawsuit was filed by some of the villagers to stop the condemnation of their homes. The lawsuit, however, was unsuccessful.

But those who moved to the Ninth Ward in New Orleans encountered a vibrant community in the early 1960s with a high rate of home ownership and community activism. The Battle Ground Baptist Church also relocated to the Ninth Ward, retaining its place in the community. But Fazendeville, the Village, all but disappeared as the land became part of the Chalmette Unit of the Jean Lafitte Unit of the National Historical Park and Preserve. Yet, through the lens of Fazendeville, and its 100-year existence, a resilient community of several generations living along the Mississippi River is revealed. Fazendeville, similar to San Malo's maroon community or the Exodusters' journey to Kansas, testified to a Mississippi River that was ever present in the lives of African Americans. Fazendeville, with a longer history than either the maroon community or the Exodusters' migration, sustained an active, self-sufficient African American community, preserving familial bonds and communal rituals such as a baptismal immersion in the Mississippi River. Adding to the river's spiritual role was the material as the area allowed for fishing, hunting, and gardening without the fear of frequent flooding. Thus, the river and its environment informed the community and memories of the Village in both material and spiritual ways.[4]

Villagers continued their relationship with the Mississippi River when many relocated to the Ninth Ward. When moving to the Ninth Ward, they were relocating to a riverine environment, carved out of cypress swamps. But unlike Fazendeville where flooding was rare, the Ninth Ward experienced floods, as the area sat beneath the levees on tracts that had been wetlands and once absorbed floodwaters and storm surges. Flooding in the Ninth only worsened after canals were built through the area. But for the first few years after Fazendeville was razed, residents enjoyed an active community where the majority were African American. The demographics of the Ninth were changing and from 1940 to 1970 as the community evolved from one that initially housed African American homeowners and immigrants, the number of non-white residents in the Ninth Ward grew from 31% to 73%. Along with the population shift, benevolent organizations—inculcating an activism within the social fabric—were part of the landscape of the Ninth Ward, also known as the Lower Nine, as residents sought to improve the services provided to the area. Home ownership was high, with many living in what was known as "shot-gun homes," which had been built with lumber harvested from nearby cypress swamps. In describing the area, Keith Ferdinand recalled, "Some of our areas were swamp lands or the rough edges of the city where you had tall grasses, marsh, trees, bushes, and shrubs, so they had to be claimed as a place to make a home." Homes were often passed down to succeeding generations, furthering a sense of belonging in the Ninth Ward. Residents were surrounded by their own stores, churches, and even a black-owned movie theater, producing what one Ninth Warder recalled, "a little world unto itself." But the riverine environment did more than supply lumber as memories of the Ninth Ward included a musical environment producing such greats as Fats Domino as well as locally well-known musicians. One local musician, Shamarr Allen, remembered how as a child "he would sit on the front porch or up on the levee for hours practicing new songs." The levee also figured in other memories as Darron Burton recalled as a child playing on the levee. In the early 1960s, employment was also linked to the river and the community. From 1918 to 1923, the Industrial Canal, also known as the Inner Harbor Navigation Canal, was built connecting Lake Pontchartrain with the Mississippi River while splitting the Lower Ninth Ward from the Upper Ninth. But the canal, which would extend for five and one-half miles, was a source of jobs by the 1930s as industrial development occurred along the canal. The canal also cut the Lower Ninth Ward off from the city center, resulting in an isolation that further fostered a sense of community among its residents. When reflecting on the Lower Ninth before Katrina, residents fondly recall family gatherings and the security that comes with lifetime neighbors. This isolation was diminished, however, when the Claiborne Avenue Bridge was built in the 1950s connecting the city and the Lower Ninth Ward and stimulating growth for the area.[5]

Everything changed in September in 1965 when Hurricane Betsy made landfall—with winds recorded at 60 miles per hour in the city and flood waters reaching 8 feet. Sixty people drowned and 7,000 homes were flooded as a result of the hurricane and subsequent storm surge. The Ninth Ward experienced the worst flooding in comparison to other New Orleans neighborhoods, in large part due to the breach in the Industrial Canal. On one block alone, all the homes were dislodged from their foundations. Reminiscent of 1927 where the most suffering occurred within the African American community, consequences for those experiencing the flooding prompted by Hurricane Betsy were disproportionate. Lower Nine residents shared their memories of those frightening days. In one account by Lucy Boyer Thomas, she recalled that on September 7,

> We knew the storm was coming because of the winds that was very high. Then what made us know also was the rain. When it was raining, and I saw the water coming up step by step. When you saw that water rising, you knew something was really wrong.

She also knew that the situation was worsening as her husband was listening to what she thought was a transistor radio. But help came for Thomas and her ten children when the National Guard picked them up and by that point, she thought the water must have reached at least 5 feet. Others were not as fortunate as Thomas talked about children drowning by the Port of Embarkation. But one memory that was shared countless times by Lower Nine residents was the scene that greeted them upon their return home. Returning home a week after the flood, Lucy Boyer Thomas recalled,

> All my furniture was floating around in here. I didn't have no furniture. I had a lot of beautiful things when I first got my . . . the silver and the chairs and all that. The kids' uniforms and everything was floating in the water. Everything was gone. Because that water was that high in the house.[6]

In another firsthand account, Ida Belle Joshua recounted the evening when the storm hit:

> But see, the storm hit us . . .It was at night when it hit us. All I could look out there and see things blowing and going on. By the time we got through negotiating about what we would do, the water was at the porch. We heard a bam, and then we saw the water.

In Joshua's interview, she also spoke of hearing an explosion during the storm, referencing what became a commonly held belief by Lower Ninth Ward residents. Many contended that the break in the Industrial Canal that caused the flooding in the Lower Ninth Ward was purposefully caused by the city—to protect white neighborhoods—and not merely a consequence of the storm. This belief had historical antecedents as the city did breach one of its levees in the 1927 Flood to protect the city. In her account, Joshua described the moment when she heard the explosion:

> Yes, we heard the explosion, but we thought it was from the storm. Then we saw all this water gushing from the levee, not from St. Bernard Parish, not where the storm was coming. The storm was coming into New Orleans, but the levee. . . when we heard the noise, the bam from the levee, the water was gushing on us from the levee.

She went on to recall a scene filled with a sense of urgency and the need to escape the rising floodwaters. The following words conveyed not only Joshua's experience but also that of her neighbors:

> I looked at my sofa that I had plastic and there was water. My husband had gotten paid that day. I say, "Oh wait! Let me go get your wallet!" He said, "Just bring your ass out of here and get in this boat!" We got in the boat, put on raincoats, with a gown and, I think, slippers or something. There children was dressed and put them in the . . .By the time we got in the boat, then the neighbors was screaming and hollering and telling him to come and get them, because they were in their homes and the water was . . . We were in a river!

Adding to the trauma of Joshua's experience was her family's efforts to escape in her husband's boat. Not only was the trip to safety threatened by floating debris, cutting the engine, but also Joshua recounted the neighbors they passed on their way to safety asking for their help. Since the boat was full, Joshua's family could not take on any more passengers and could only promise to send help.[7]

Other memories of Hurricane Betsy persist, but along with the accounts of drowning and harrowing experiences of flight were recollections of rebuilding the community. One institution that once again faced the prospect of starting over was the Battle Ground Baptist Church. But the church was rebuilt as others were and unlike the devastation that was experienced with Katrina, after Betsy, Lower Nine residents claimed, "activism was heightened." Of course, damages resulting from Hurricane Betsy were less than Katrina as Betsy's loss amounted to $6.5 billion in 1990 compared with Katrina with losses of $82.2 billion. In 1965, however, Betsy was devastating for New Orleans, especially the Lower Ninth Ward, prompting what some African Americans believed was the "white flight" to St. Bernard Parish. The statistics supported their contentions. By 2000, the Lower Ninth Ward had a population base that was 90% African American and among the population, a third lived in poverty. Adding to the perception that Betsy triggered the exodus of white residents from the Lower Ninth Ward was another narrative that framed the disaster for African Americans. The majority of white residents understood Hurricane Betsy as a natural disaster, whereas African Americans identified the disaster as man-made. The contention that the canal was dynamited to safeguard white neighborhoods framed Betsy memories, reinforcing a history of the river that differed significantly from the dominant narrative. This counternarrative, which would be reaffirmed with Katrina, spoke to an African American history with the Mississippi rife with opposites as the river imprisoned—through an infrastructure laden by racism—yet laid the seeds for rebirth. But by the dawn of the twenty-first century, the Lower Ninth Ward was still intact, a far different community than the one that emerged from the wreckage of Hurricane Katrina.[8]

The damage wrought by Hurricane Katrina was unprecedented as the storm and the resulting storm surge resulted in hundreds of lost lives whereas thousands of others were left homeless, jobless, and often traumatized as they coped with new environments. The events leading up to Hurricane Katrina began in the last days of August 2005 when the storm was identified as a category 3 hurricane that was beginning to bear down on the Gulf Coast. Initial responses to the storm included the evacuation of almost one million people from the region, with an estimated 70% leaving.[9] But the problem of waiting out the storm remained for more than 100,000 people who were without access to transportation, particularly in areas such as the low-lying Ninth Ward. Adding to the challenge of transportation, was the lack of funds especially as Katrina occurred at the end of the month when those who relied upon monthly checks, were without funds. Thus, many were without resources when the surge waters of Katrina hit. So although there were warnings to evacuate—and this also became an issue as some held Mayor Roy Nagin and ineffectual leadership responsible for the needless suffering that ensued—many were simply without the means to leave and without the resources to find another place. To explain this even further, when Katrina hit, one in four African Americans earned less than $20,000 annually. (This is even more significant because one of the neighborhoods in New Orleans most damaged by the Katrina storm surge had a predominantly African American population.) The same statistics hold for transportation: One in four African Americans did not own a car. So, when Katrina made landfall, many not only did not own the necessary transportation that would have allowed them to flee from the storm's path, but they also did not have the safety net of additional resources to support evacuation. As a result, another Katrina calamity was the evacuation of an estimated 30,000 residents to the Superdome in hot, humid New Orleans weather. The scenes and stories from those forced to wait out the storm are retold in dozens of firsthand accounts as they recall failed plumbing, part of the roof blowing off, an inefficient food supply network, and a general feeling of lawlessness. In one memoir by Paul Harris, he recalled, "the stench of backed-up toilets, children sleeping on cardboard soaked with urine and feces, 90-degree heat and oppressive humidity after part of the roof blew off and the electricity failed." Another survivor compared staying at the Convention Center with African Americans forced to remain on the levees in the 1927 Flood. At the Convention Center, six deaths were recorded with one man throwing himself off a balcony, another dying from a drug overdose, and four others dying from natural causes.[10]

In the sequence of events after Hurricane Katrina making landfall, initially New Orleans' residents were relieved that the storm had not caused significant damage—the levees had held (Figure 5.2). But the relief was brief as surge waters from the hurricane began to flood the city. The massive devastation that resulted came from the storm surge, as the surge that followed was the largest one ever to be recorded in U.S. history. On the Gulf Coast, there was a surge of about 30 feet high covering a 20-mile radius. New Orleans would see a storm surge of up to 19 feet high in places. Not surprisingly, the destruction was unparalleled, even for a city used to hurricanes. Several levees collapsed, and other levees would be eroded as the water poured over the tops. In the past, before the construction of canals through the marshlands and the development of neighborhoods such as the Ninth Ward, the wetlands absorbed the overflow and acted as a buffer. But what made Katrina even more horrific but predictable—given the geography of areas such as the Ninth Ward—was the disproportionate damage suffered by the African American community and the indifference by federal officials that followed. According to one New Orleans' expert, Ari Kelman, "In the city proper 80% of flooded neighborhoods were majority nonwhite." Other Katrina statistics included the number of deaths as an estimated 1833 people lost their lives. In addition to the official count, recent scholarship has noted the number of stillborn deaths due to the stresses of Katrina, followed by Hurricane Rita a month later. Scholars place this number between 117 and 205 additional deaths. Other telling statistics include that a land mass of 90,000 square miles was inundated along with the destruction of 300,000 homes. As mentioned earlier in a comparison with Hurricane Betsy, the economic loss was an estimated $125–$150 billion. One of the saddest facts was the number of bodies either unclaimed or unidentified, totaling 200. Adding to the trauma of losing a loved one was the way many died, with survivors unable to bury the bodies for extended periods. Compounding the loss of loved ones was the imagery conveyed by the media with visuals of the dead floating among the streets in New Orleans. This visual, in particular, became part of the Katrina landscape.[11]

FIGURE 5.2 Survivors Walking through Floodwaters.
Credit: Historic Disasters—Hurricane Katrina, FEMA.gov.

Despite the composite picture of images conveying loss and despair, this history with the river was not new for African Americans. Similar to the Mississippi Flood in 1927, again the damage was disproportionate. A brief overview from the preceding chapter will illustrate the similarities. While the official number of fatalities from the 1927 Flood was 250, more recent scholars estimated that the death toll could have reached 1,000. Of those 1,000, at least 100 African American men drowned on one night alone as National Guard troops forced them to remain at the Mounds Bayou levee in a last-ditch effort to save the levee. This levee, outside of Greenville, Mississippi, along with more than 200 others, failed with floodwaters covering an area 70 miles wide, inundating more than 26,000 square miles in seven states, leaving almost 700,000 to one million without homes and property damage reaching the billion-dollar mark. Of those displaced, the majority (90%) were African Americans, revealing a world where Jim Crow laws prevailed and peonage was the norm on the last remaining Southern cotton plantations. In the midst of multiple flood narratives, the experiences of African Americans living in the Mississippi-Yazoo Delta are retold, emphasizing their harsh, often brutal mistreatment by Southern whites—witnessed by indifferent federal and aid officials—resulting in a mass migration to urban areas in the north and south. Finally, the similarities between the responses of the federal government and relief agencies in 1927 and Katrina are numerous. For example, in 1927, the Red Cross often relied upon white landowners to distribute aid to African American survivors, often resulting in a corrupt system that further curtailed African American efforts to start over. When the federal government began to award funding to rebuild New Orleans, residents in the Ninth Ward were often the last to receive support, leaving their neighborhood one of the last to be resettled. (Also, when federal aid was distributed, it was not enough to rebuild homes.) Yet when Katrina hit, the Ninth Ward experienced some of the worst flooding as it was situated on the low-lying lands. African Americans had experienced this geographical vulnerability earlier when during the 1927 Flood, black tenant farmers also were relegated to the bottomlands, the first lands to be flooded. But there were differences and what made Katrina worse was the additional 100 years of riverine engineering that resulted in the loss of wetlands—carved up by oil and gas activity—which had served as buffers for hurricanes. Adding to the disruptions of the wetlands' ecological regime was the construction of levees that reshaped the Mississippi River. Compounding Katrina's excesses was an inadequate infrastructure for transportation and emergency relief. All of this would become part of a brief national conversation.[12]

One constant, however, was the response of the African American community. Similar to the 1927 Flood, Katrina would be in part defined through the cultural production of song, prose, poetry, and art. Almost immediately, musicians responded to the disaster, some well known such as the rapper Lil' Wayne, a New Orleans native, who evoked memories of earlier responses by blues artists, such as Charley Patton, Kansas Joe, or Memphis Minnie, and their flood blues songs framing the 1927 Flood. In "Tie My Hands," Lil' Wayne captured the frustration many felt toward the federal government as well as the despair as visuals of floating bodies persisted. His lyrics cite a chronology of events, first the hurricane, followed by the morning sun. Within the lyrics, references are made to the dead while the city of New Orleans is submerged under unforgiving flood waters. Events continue to unfold with allusions to a president "still choking," pointing to the slow response of President George W. Bush to the hurricane's devastation and the insulting label of "refugee." In "Tie My Hands," Lil' Wayne exposes the hypocrisy of the label as the survivors were born in the United States. Lil' Wayne's lyrics mirror the frustration that Charley Patton voiced in his 1927 Flood ballad, "High Water Everywhere" as Patton conveyed the sense of ever-present floodwaters paired with the impossibility of flight to the hills.[13]

Another example of popular African American musicians responding to the tragedy of Katrina was the Houston-based hip-hop group, K-Otix. The group composed a rap song centered around Kanye West's observation that George Bush did not care about black people. Using West's comments that were spoken during an interview soon after the events of Katrina unfolded, the song was titled, "George Bush Doesn't Care About Black People." Similar to Lil' Wayne, the song conjured up images and news reports that had been dominating the media. For example, the song references the survivors who were trapped in their attics—some for five days—without access to their cell phones. Another commonly heard theme cited by K-Otix relayed the inept rescue efforts recounted by survivors, such as one story of helicopters rescuing some off their rooftops but not returning for those remaining, despite promises to come back. And of course, the group reminds listeners that those left behind were without means to evacuate—a circumstance that echoed the experiences of those living through the 1927 Flood. One of the most damming critiques is the group's references to the Federal Emergency Management Agency (FEMA) as K-Otix states, "If FEMA really comes through in an Emergency, But nobody seems to have a sense of Urgency."[14]

Other musical responses to Katrina and particularly from one of the genres that evolved out of hip-hop but with its own New Orleans spin are called "bounce." After a historical trajectory dating back to the time of slavery, music scholars have noted how this type of hip-hop has its antecedents in the "call and response" that was initially part of the work songs of enslaved African Americans. Using this approach, it was often based on the surrounding community and after Katrina, New Orleans scholars still found traces of it being used in some of the displaced communities. In addition to the well-publicized musical responses to Katrina, a new medium emerged with amateurs crafting their own songs using the YouTube platform. Self-made artists crafted videos such as one homemade video, titled "Hurricane Katrina and Rap Song." The artist included moving images—the Convention Center, people waiting on rooftops, residents wading through floodwaters holding their children, and helicopters circling above. Along with images, amateur videos included the themes associated with Katrina—the high temperatures after the storm increasing the suffering of children and others, often without food or water; the indifference of federal officials, causing so many deaths to occur; and of course, the racism that prompted the disproportionate suffering by the African American community.[15] Regardless of whether a well-recognized musician or an amateur responding to this latest Mississippi River inundation, a consistency with earlier Mississippi River floods and the African American community is revealed. This cultural production is distinct and unique to the African American community that lived along the Mississippi River. In contrast to other communities living along the Mississippi, the themes reflect not only a persistence and endurance, but also an intimacy with the river and the devastation it can cause.

Poets also added to the chorus of loss, persistence, and endurance. One poet who successfully conveyed the continuity of African American history in the United States is Claire Crew and her poem, "Sitting Ducks at the Superdome." The poem is lengthy, with a few stanzas provided below:

"Sitting Ducks at the Superdome"

One day in America her dirty secrets were revealed
Hurricane Katrina, Neglect and the Levee broke

New Orleans August 2005
God bless America
African Americans herded into a superdome,
Lying, dying in darkness waiting to be rescued

Waiting for the descendants
Who captured their ancestors into slavery,
Forcing them in shackles ashore
On the waters to the Americas,
Cramped tight on ships where many died

Of lack of food and water.
Waiting they sit in the super dome,
Cramped tight dying of lack and food and water.
Surrounded by the dirty sewer
Swamp waters of New Orleans[16]

And still another poet responded to Katrina. Yusef Komunyakaa, in his poem, "Requiem," like Claire Crew, looked to history to frame Katrina and its consequences:

"Requiem"

So,
when the strong unholy high winds
whiplashed over the sold-off marshlands
eaten back to a sigh of saltwater,
the Crescent City was already shook down to her pilings,
her floating ribs, her spleen & backbone,
left trembling in her Old World facades
& postmodern lethargy, lost to waterlogged
memories & quitclaim deeds,
exposed for all eyes, damnable
gaze & lamentation—plumb line
& heartthrob, ballast & watertable—
already the last ghost song
of the Choctaw & the Chickasaw
was long gone, no more than a drunken curse
among the oak & sweet gum leaves, a tally
of broken treaties & absences echoing
cries of birds over the barrier islands
inherited by the remittance man, scalawag,
& King Cotton, & already the sky was falling in on itself,
calling like a cloud of seagulls
gone ravenous as the Gulf
reclaiming its ebb & flowchart

while the wind banged on shutters
& unhinged doors from their frames
& unshingled the low-ridged roofs
while the believers hummed
"Precious Lord" & "Deep River"[17]

Complementing the cultural production of musicians and poets were the artists who contributed to the cultural chronicling of Katrina, leaving another lens in which to understand the aftermath (Figure 5.3). One of the most compelling paintings was conceived by Ted Ellis, an African American artist who grew up in New Orleans. In his work, "Surviving Katrina," an African American family is depicted stranded on their roof, surrounded by flood waters while the father is erecting the American flag. Although stranded, however, the family does not convey the imagery of hopelessness that John Steuart Curry depicted in his 1937 painting, "The Mississippi." In Curry's work, the father is beseeching the heavens with his wife and daughter behind him on the rooftop. They are also surrounded by floodwaters in what appeared to be a situation that will only be rectified by intervention from a higher power. The contrast between the two works represents differing memories—Ellis framing the African American experience with Katrina as one of survival and mastery and Curry shaping a discourse of victimhood. Supporting and featuring the work of Ellis is the Lower Ninth Living Museum, founded in 2011. The museum that included other Ellis' paintings, such as "Born by the Bayou"—a landscape scene—was a product of

FIGURE 5.3 Superdome Katrina.
Credit: Photo by Nuri Valbona, Abaca Press/Alamy stock photo.

the Ninth Ward's slow recovery. The museum is committed to "remembering and celebrating the vibrant history of this neighborhood."[18]

Adding to the understanding of Katrina in the context of an African American history of the Mississippi River are the numerous personal accounts. Similar to earlier flood narratives, Katrina memories followed the same tropes of displacement, loss, and ill treatment by relief groups. In addition to these constants were the generational strategies for surviving a flood shared by survivors. Distinguishing Katrina, however, are all the contemporary media outlets, shaping numerous Katrina accounts in comparison to the 1927 Flood and to a lesser degree, Hurricane Betsy. With the attention that Katrina received from various venues, the result was a wake-up call for many as to the plight of African Americans living in the flood-prone, low-lying lands. Yet while reports of post-Katrina conditions were not whitewashed by an aspiring presidential candidate, such as the actions of Herbert Hoover in the 1927 Flood, a systemic racism dominated news reporting. Images of looting and theft by African Americans frequented the news, whereas the differential treatment afforded to white and African American survivors did not receive the same coverage. Use of the word "refugee" by journalists to describe African American survivors was soon attacked as African Americans perceived the term denigrating, obscuring their status as citizens. So, for African Americans reflecting on their history with the river dating back to the 1700s, their negotiations with a riverine environment dominated by engineers, bureaucrats, and policy makers could boast of a consistency fostered by race and class and resulting in differentiated treatment and outcomes. But another and an equally important constant lies within the resilience found in the African American community, such as that expressed by the pastor of the Battle Ground Baptist Church or others in the Lower Ninth Ward who were determined to return and rebuild. The commitment to place was again evident as seen after Betsy. Interviews with New Orleans' residents illustrate these frustrations as well as the determination to persevere as in earlier floods.

In reviewing several firsthand Katrina experiences, one survivor revealed how coping with Mississippi River Floods was passed down through generations. His memory evoked 1927 memories of survivor accounts of how they saved their lives and homes. In his words, his father taught him to always ". . . carry an axe and a hatchet. If a hurricane came, make sure you had water, canned goods and stuff like that." He went on to say, "My father really schooled me on hurricanes." Complementing this memory was that of others who decided not to evacuate, citing earlier experiences with hurricanes. For example, in a study of 53 African American men who chose to remain in New Orleans, one participant remarked, "course it's always been that way with us. I have stayed through many storms, even through Hurricane Betsy. But the storm would come through, we have our flood and get back on track." Other comments included, "If I survived Hurricane Betsy, I can survive that one, too. We all ride the hurricanes, you know." Finally, the familiarity with earlier hurricanes and earlier evacuation orders, such as the one for Hurricane Ivan that resulted in long, exhausting trips to nearby cities, cultivated in some New Orleanians a wariness about evacuation orders.[19]

Still another account conjured up a familiar flood scene, namely survivors waiting on their rooftops for days. Leonard Smith, who before Katrina was very active in the community, shared his experience as floodwaters reached his home in the eastern part of New Orleans:

> I had some things I was going to take up in the attic, but I didn't have time to get them. I didn't have my keys and my glasses or nothing. By the time I got to the ladder to get to the attic, the water just exploded. And everything just went straight up in there: the couch, the TVs, the refrigerator, and the freezer. . . . Now I'm in the attic and the water's rising so fast. . . . For the next three days, I was in and out of the attic on the roof. You had to go on the roof when the helicopters were comping, so you can see if they can see you.

Finally, one of the helicopters picked Smith up and took him to a staging area where he was given water and oranges. Later, Smith was moved to a shelter in Houma, Louisiana where his family found him. Despite the three days of uncertainty, compounded by a medical condition, Smith planned on returning to New Orleans. In other personal accounts, survivors echoed the same suspicions that surrounded Hurricane Betsy, with many believing the levees were breached purposefully, again sacrificing African American neighborhoods.[20] When interviewing Pete Stevenson who lived in the Upper Ninth Ward when Katrina hit but since then had moved permanently to Cullman, Alabama, he not only did not plan to return to New Orleans but also had this to say about the hurricane. "The blowing up of the levees was meant to kill the blacks and the poor whites." In another interview with Parnell Herbert, he echoed Stevenson's thoughts, "I'm not going back to New Orleans for them to allow something like this to happen." "I believe there was a deliberateness to allow tactical portions of the levees to deteriorate so that they would be the weak points. They couldn't beat us otherwise. So that's how they drove us out." In numerous accounts, survivors insisted that not only did the government allow for the levees to become substandard and weakened but that they were also blown up as in past hurricanes.[21]

Regardless of whether canals were blown up or breached by the overwhelming floodwaters, everyone identified with the suffering that so many underwent. In one of the most moving accounts, Robert Green recalled how he and his brother, Jon, tried to save his mother, a cousin, and three granddaughters. His mother, who was bedridden, would not make it to safety nor did one of his granddaughters. Green, in reliving the experience with a local journalist, remembered flood waters entering their home and as the house began to break up, he attempted to lift all the children to the only safe place, the roof. The following excerpt recalled the next few minutes:

> Robert reached for the first child he could get his hands on, 'Nai-Nai' and lifted her up. As he turned for the next, he heard a splash behind him.

> He turned, and 'Nai-Nai' was gone. He froze in horror. He screamed to Jesus. He hesitated, he cursed. Robert wept as he tried to rationalize his next move—a response he's relived over and over for the past year. The house crumbled. In a split second he decided to continue rescuing the other children.

Unfortunately, Green's story was not an isolated one. For others, such as Denise Roubion-Johnson, while her family's experience was not the tragedy that Green experienced, they still endured horrific conditions during the surge and relocation. For Roubion-Johnson, the decision to remain in New Orleans and ride out the storm was governed by her husband's health issues. Since Roubion-Johnson was a nurse, she volunteered to work at University Hospital during the storm, provided there would be a place for her husband and teenage son. When the family arrived at the hospital, they initially believed, like many others, that the storm had passed but by Monday night, water started coming in the hospital. From Monday evening until the following Saturday, Roubion-Johnson and the others were trapped inside with no power or running water. In a very graphic description, Roubion-Johnson described a scene testifying to the trauma many must have experienced:

> Things you take for granted like lights, air conditioning, and ventilators for The Intensive Care Unit: There was none of that. God, it was stifling hot. We were going to the bathroom in plastic bags. Everybody: patients, nurses, doctors. The smell of the feces and urine was unbelievable. It got worse and worse. There was no running water after a few days, so there was no bathing. You'd just wipe off with normal saline.

Unfortunately, after leaving the hospital, Roubion-Johnson's experiences did not improve for several weeks. First, in the midst of securing safety for the patients, Roubion-Johnson was separated from her family. Second, as the military escorted busloads to several stops in Texas, Roubion-Johnson said they "were herded and treated like cattle" and compared the experience with going to a concentration camp. Insulting her further was the reference to her as a "Katrina refugee" since she did not have any ID with her. In Roubion-Johnson's words, "Regardless of our socioeconomical status, we were Americans!" Still, Roubion-Johnson and her husband returned to New Orleans and hoped to eventually have her family reunited in the city.[22]

Others shared this desire to return, despite the hardships they underwent. For example, Joe Bridges, an African American contractor doing electrical work in New Orleans, shared his post-Katrina story. While Bridges experienced the flood differently than earlier survivors in the 1927 Flood or even Hurricane Betsy, parallels remained, such as the attachment that Bridges felt to New Orleans. This sense of place prompted Bridges to return to New Orleans and help rebuild the community. In Bridges' story, when the flooding first occurred, he and his family were forced to move, ending up in Atlanta and later Washington, D.C. After the move to the nation's capital, Bridges decided to commute to New Orleans, work for one to two weeks, then rejoin his family in the Washington, D.C. for a few weeks, followed by a return to New Orleans. Despite opportunities in the nation's capital, Bridges felt compelled to return and in his words,

> I started thinking about it, and I'm like, You do electrical work, and three quarters to seven eighths of your city is out of power and you're just gonna leave, and just leave your city hanging? The place that helped you make your money, that you grew up, that you were born here?

Bridges and his family did return in January 2006, rebuilt the family home, with family members still living there today. His experience demonstrated again the resilience that has been part of the fabric of existence for African Americans living along the Mississippi River.[23]

Bridges, however, is not an anomaly as many insisted on returning to the area they considered home. In one story, Shamarr Allen, a young musician, not only moved back to New Orleans, where he grew up in the Lower Ninth Ward, but also offered free music lessons to aspiring musicians in the Ninth Ward. He remembers as a child, "sitting on the front porch or up on the levee for hours practicing new songs." Other accounts include the return and rebuilding of the Battle Ground Baptist Church, as the Reverend Sanders reiterated the importance of the church and his hopes for rebuilding:

> If we can build Battle Ground back, the church, it will mean a lot. The church is scattered right now but they love BGBC and if the church is rebuilt, the people will come back. . . . But, I believe with a genuine effort, with the right people behind us, I think it (the return) can happen.

For some the return to New Orleans, such as those coming back to the Ninth Ward, was motivated by a sense of place that identified with the unique Mississippi River environment. In a 2006 survey of 103 former New Orleans residents displaced in Houston neighborhoods, 55 preferred life in New Orleans, citing lifestyles that included enjoying the bounty of the Mississippi Delta with barbecues, eating crawfish and shrimp. Accounts also referenced the openness of New Orleans society, reflecting the "edginess" that another New Orleans survivor associated with the Lower Ninth Ward.[24]

Not all survivor stories, however, concluded with a return to New Orleans. Many flood victims were permanently displaced to cities such as Houston, Texas, or Atlanta, Georgia, replicating the Northern migration after the 1927 Flood. Their displacement reflected the complete loss of home and livelihoods and sparked an outpouring of news reports as residents attempted to negotiate this new environment. But for some, the move to other cities also resulted in a better life as former residents chose to remain. For those who stayed in cities, such as Houston, their reasons included the availability of more amenities. Better schools, jobs, and medical care were all cited as reasons to stay and not return to New Orleans. Again, the river proved to be a catalyst for movement—provoking some to return and start over whereas others recognized the flood as the push to start a better life elsewhere.[25]

For those, however, who wished to return, the city itself was not completely dewatered for six weeks. Worsening the situation was the arrival of Hurricane Rita, which extended the recovery time. Compounding the physical deterrents to returning were the actions or, in many instances, the inaction of the federal government. This became a major grievance for the African American community, as the financing provided for housing was not sufficient to rebuild in the Ninth Ward where home ownership had been high with more than half owning their own homes. Many considered the housing crisis the "second disaster." And the statistics support the inequities that the city's African American population had decreased by 28% by 2017 compared with a decrease of 7% of the white population. In the Ninth Ward, by 2018, the population was still only one-third of what it had been before Katrina with fewer than 5,000 residents. Still growth was occurring, which testified to the determination of those intending to return. By 2020, there were 1,675 households compared with only 1,060 in 2010. Before Katrina, there were 4,820 households recorded in 2000. Still, these numbers should be framed within the greater New Orleans rebound where 90% of the population had returned.[26] But whether to return or permanently move away, a sense of loss attended all of those whose homes were flooded. Keith C. Ferdinand best summed up the experience of loss when he said,

> You're not mourning the loss of your '76 Buick. You're mourning the loss of friends and colleagues who may have died or been crippled, the everyday common things like the store you went to, the church you attend, and the gas station in New Orleans East that has been leveled. You mourn the loss of the city, your sense of your neighborhood.

Richard Mizelle, Jr., a contemporary historian, added to Ferdinand's insights when he asked readers to consider the loss of intangibles. In his words, consider loss such as a "family Bible, that recorded your baptism as a child, the pocket watch your father gave you, or the countless pictures of relatives lining the wall or in photo albums that are part of your existence." These are the losses that become etched in the collective memory of a river that not only sustains lives but also diminishes in the abstract and physical sense.[27]

By the time Katrina hit the Gulf Coast, African Americans' history of the Mississippi River had changed substantially since the early 1770s (Figure 5.4). The relationship between the river and the community was now fraught with layers of policy generated by multiple bureaucracies, entrenched in centuries of indifference. Beginning with the 1927 Flood, new actors were directing the river with hopes of containing or at the least subduing the Mississippi's will to flood,

FIGURE 5.4 Katrina's Commemorative Second Line.
Credit: Historic Disasters—Hurricane Katrina, FEMA.gov.

finding new channels in its pilgrimage to the Gulf. While commercial interests prompted the actions of engineers and policy makers in their attempts to contain the river, there was still space for multiple river narratives. Despite radical change such as the erasure of Fazendeville and threats of extinction to neighborhoods, such as the Ninth Ward with dislocation for so many as major flooding threatened the return of its residents, community members did return. In a 2011 interview with John Taylor, an African American man in his sixties, he discussed how he spends his days in an area of the Lower Ninth Ward overlooking the Bayou Bienvenue, where he acts as a guide for people visiting the bayou. Since he grew up in the Ninth Ward, he remembered Hurricane Betsy and the geography of the area in 1965. As a child before Hurricane Betsy, the area was bountiful as he fished "among the baldycypresses." He described a rich ecological terrain as he recalled, "Back then you couldn't even see across to the other side of the bayou because the woods were so dense. And the water was covered with lily pads." Taylor was also aware of the functional nature of the marshland cypresses. He related how they absorbed floodwaters and speculated about their demise with the construction of the Industrial Canal, ensuring the destruction that came with Betsy and Katrina.[28]

Still the clash of memories—built upon centuries of lives that intersected with the river through bondage, through escape, and through building homes, urban and rural—with the actions of policy makers, corporate interests, and the engineering community mark the future of New Orleans. Underlying these new threats were the actions of policy makers, willing to sacrifice these areas with their long-standing community ties for areas considered more prosperous and instrumental to the future of New Orleans. Yet, there were pockets of resistance; for example, the work of churches such as the Battle Ground Baptist Church and the commitment of many to return to the city. Of course, this chapter only explored the Lower Mississippi Valley and the area around New Orleans. But adding to this new chapter with the river and its environmental woes, including the loss of wetlands the size of one football field every 100 minutes, was the despoiling of the nation's major waterway through the petrochemical factories lining the river.

By the time the floodwaters of Katrina had subsided, multiple narratives were emerging, assessing the damage and the aftermath from various perspectives. But for African Americans, one narrative predominated over the rest—the federal government's role in the suffering that resulted from Katrina. From the indifference that began with President George W. Bush's reaction to the slow recovery of African American neighborhoods in comparison to white communities to the unshakeable belief that the Industrial Canal was blown up to save other neighborhoods—all of these contributed to an ongoing narrative of African Americans and the Mississippi River that has been unchanging. Journalists and scholars joined in the criticism, such as *New York Times* columnist Paul Krugman, who, in his assessment of Katrina, critiqued the "cronyism and incompetence of President George W. Bush administration." The historian Douglas Brinkley was unstinting in his criticism of New Orleans mayor Nagin and how he mishandled the evacuation. Other scholars, such as Karen O'Neill, found fault in the city's infrastructure where public works and emergency planning that would have aided poorer communities were lacking. The chorus of support for the African American community, recognizing the structural inequalities that have resulted in such unfair and uneven treatment, reflected what for African Americans was a historical relationship with the river. For the residents of the Ninth Ward or earlier 1927 Flood survivors, their relationship with the river was characterized by its consistency. The river has always been viewed not only as a catalyst for despair and loss but also a refuge and in this instance, the magnet, sustaining a sense of place.[29]

Another narrative concerned the river itself. By the time of Katrina, the river's decline in status was complete as engineers and policy makers in their modernizing zeal reconfigured a river that once was as much feared as revered. For African Americans after Katrina, while the outcomes for many were the same as those that followed the 1927 Flood, the rhetoric and reactions differed. For example, when analyzing the lyrics in flood blues, in many cases, the artist recognized the agency of the river. As discussed earlier, in one of Bessie Smith's songs, "Homeless Blues," she lamented the cruelty of the river when she sang, "Mississippi River, what a fix you left me in." She continued with blaming the river for the deaths of her parents and concluded, "Mississippi river, I can't stand to hear your name." While other artists, such as Charley Patton, recognized the discriminatory practices of not letting African Americans leave the area, or DuBois urged African Americans to flee the Delta, still the flooding and initial suffering was attributed (in most instances) to the river. With Katrina, the riverine environment is seldom mentioned although one survivor, Evangela Bailey, whose family was forced to relocate to Houston asked her new minister to rebaptize each family member and "to wash off the stain of the angry waters." For most survivors, however, focus remained on the indifference of the power structure whether in the responsibility for dynamiting a levee or the differential treatment in evacuation, for example. Yet the loss of neighborhoods such as the Ninth Ward are recognized for the sense of place they offered with references to the swamplands or the lifestyle of fishing and gardening that the environment allowed. Further, in African American discussions of Katrina, scholars consider the similarities between the causes and outcomes with those resulting from the 1927 Flood. Even in firsthand accounts, the volatility of this riverine environment is missing from the discussion. Yet, in accounts such as the one by Keith C. Ferdinand, he acknowledged how the environment of the Ninth Ward with its swamplands—considered unattractive by most—shaped their lives, giving them an edginess, pioneering spirt, as well as the opportunity for home ownership. So, by the late twentieth and twenty-first century, the discourse with the river was changing. The riverine environment became the site for homes, framing lifestyles and informing a culture that became unique to the river. These nuances remained in other communities along the Mississippi River as the river became polluted, which, in turn, contaminated African American neighborhoods along the waterway; again, this occurred in neighborhoods where home ownership was high. The Epilogue will review these changes, as the intersection of African American lives with the Mississippi River reflects contemporary concerns of climate change and its effects.[30]

Notes

1 For general histories of Fazendeville, see Joyce Marie Jackson, *Life in the Village: A Cultural Memory of the Fazendeville Community* (Washington: U.S. Department of the Interior, National Historical Park, 2003); Jackson, "Declaration of Taking Twice: The Fazendeville Community of the Lower Ninth Ward Empowering Place" *American Anthropologist*, Vol. 108, No. 4 (December 2006):766–780; Allison H. Pena, "Wade in the Water: Personal Reflections on a Storm, a People, and a National Park" *American Anthropologist*, Vol. 108, No. 4 (December 2006):781–798; Robert W. Blythe, *Administrative History of Jean Lafitte National Historical Park and Preserve* (Washington: U.S. Department of the Interior, National Park Service, 2012); April Antonellis, "New Orleans and Fazendeville (de)Segregated: Challenging a Narrative of School Integration" M.A. Thesis (Louisiana State University, 2013) accessed at https://digitalcommons.lsu.edu/gradschool_theses/2346/; Eve Abrams, "The Defeat of Fazendeville" *64 Parishes* (Louisiana Endowment for the Humanities, n.d.) accessed at https://64parishes.org/defeat-fazendeville

2 Craig E. Colten, "Vulnerability and Place: Flat Land and Uneven Rise in New Orleans" *American Anthropologist*, Vol. 108, No. 4 (December 2006):731–734. Numerous histories of the Ninth Ward are available. One of the more comprehensive ones is authored by Alexandra Giancarlo, "The Lower Ninth Ward: Resistance, Recovery,

and Renewal" M.A. Thesis, (Louisiana State University, 2011) accessed at https://digitalcommons.lsu.edu/gradschool_theses/1439/ In her overview, she utilized the subdisciplines of critical race theory, resistance theory, and social memory in explaining the Ninth Ward's successes and challenges.

3 Ron Chapman, "Fazendeville" *MyNewOrleans,com*, 28 December 2004 accessed at https://www.myneworleans.com/fazendeville/; Jackson, *Life in the Village: A Cultural Memory of the Fazendeville Community*, p. 39 as quoted in Pena, "Wade in the Water," p. 782. In the same article, the claim is made that Fazendeville had only 45 families at the most. Pena, p. 778.

4 "The Story of Fazendeville: An African American Town in the Middle of a Battlefield," accessed at wgno.com/news-with-a-twist/culture/the-story-of-fazendeville-an-african-american-town-in-the-middle-of-a-battlefield/ 20 April 2021. According to this article, the town was uprooted "to keep blacks from voting and school districts at a time when the county was at the precipice of desegregation." Blythe, *Administrative History of Jean Lafitte National Historical Park and Preserve*, pp. 74–76. Although the majority of Fazendeville residents were vehemently opposed to the destruction of their community, there were accounts from those not unhappy with the forced move. In Rose Cager's words, "All them little raggedy houses in the Village, people were glad to get in some decent houses. I don't know nobody that was angry." Quoted in Blythe, *Administrative History of Jean Lafitte National Historical Park and Preserve*, p. 76. In Giancarlo's thesis, she also referenced Fazendeville residents who had left the community before its condemnation in search of housing with more area amenities. See Giancarlo, "The Lower Ninth Ward" (2013). One of the most criticized aspects of the Fazendeville condemnation was the inadequate compensation awarded to the residents. Not only were properties in St. Bernard Parish selling for higher prices but the state median value was also higher at $10,700 in 1960. See Blythe, *Administrative History of Jean Lafitte National Historical Park and Preserve*, p. 96. Not only were the homes in St. Bernard valued at higher prices than those in Fazendeville but in 1960, the Louisiana State median home value was $10,700, p. 96; Sushant S. Mukherjee, "On a Battleground in New Orleans, A Lost African American Village Fades from Memory" 14 February 2019, maybe accessed at https://sushantsmukherjee.com/2019/02/14/on-a-battleground-in-new-orleans-a-lost-african-american-village-fades-from-memory/

5 Juliette Landphair, ""The Forgotten People of New Orleans": Community, Vulnerability, and the Lower Ninth Ward" *Journal of American History*, Vol. 94 (December 2007):837–845 accessed at http://archive.oah.org/special-issues/katrina/Landphaira57a.html?link_id=dev_9thflood; According to one New Orleans scholar, the city as a whole is "one of the most stable, multigenerational urban populations in the country." See Shannon Lee Dawdy, *Patina: A Profane Archaeology* (Chicago: University of Chicago Press, 2016):5 Keith Ferdinand Interview, in *Overcoming Katrina: African American Voices from the Crescent and beyond*, ed. D'Ann R. Penner and Keith C. Ferdinand (New York: Palgrave MacMillan, 2009): 92; *Washington Post*, 28 August 2015. In documenting the evolution of the Ninth Ward, Landphair provides another interesting statistic in that the nonwhite population of the Ninth Ward grew from 31% in 1940 to 73% by 1970, five years after Betsy. For a historical overview of the Ninth Ward, see Giancarlo, "The Lower Ninth Ward."

6 Craig E. Colten, *An Unnatural Metropolis: Wresting New Orleans from Nature* (Baton Rouge: Louisiana State University Press, 2005). In this work, Colten analyzed the geography of the Lower Mississippi Valley, explaining how flood-prone areas—built on low-lying lands—such as the Ninth Ward would suffer disproportionately from hurricanes and tidal surges such as the one that occurred with the breaching of the Industrial Canal. In explaining why Hurricane Betsy so affected Ninth Ward residents, he wrote, "such were consequences of development that relied on structural protection," p. 146; Andy Horowitz, "Hurricane Betsy and the Politics of Disaster in New Orleans" *Journal of Southern History*, Vol. 80, No. 4 (November

2014):893–934; *Washington Post,* 28 August 2015; Lucy Boyer, interview by Nilima Mwendo, 21 November 2003, interview 4700.1687, transcript, Louisiana and Lower Mississippi Valley Collections, Louisiana State University Libraries (hereafter referred to as LLMVC).

7 Ida Joshua, interview by Nilima Mwendo, 20 November 2003, interview 4700.1684, transcript, LLMVC; Multiple firsthand accounts either attribute or suspect that the Industrial Canal breach during Hurricane Betsy was caused by city officials. See Kalamu ya Salaam Interview, Keith C. Ferdinand Interview in *Overcoming Katrina,* pp. 83, 99. After Katrina, some contended that the levees were again purposefully blown up. In the words of one Katrina survivor, he shared, "The blowing up of the levees was meant to kill the blacks and the poor whites." See Pete Stevenson Interview in *Overcoming Katrina,* p. 34. Contradicting these contentions is the work of Edward F. Haas, "Don't Believe Any False Rumors . . ." Mayor Victor H. Schiro, "Hurricane Betsy and Urban Myths" *Louisiana History,* Vol. 45, No. 4 (Autumn 2004):463–468. Haas contends this was not the case but the belief that the city dynamited the canal has become the stuff of urban legends. For a broader discussion of the counternarratives regarding Hurricane Betsy, consult Horowitz, *Katrina: A History, 1915–2015* (Cambridge: Harvard University Press, 2020):44–69.

8 Deon Roberts, "Hurricane Betsy Aid was Miniscule Compared with Katrina" *New Orleans City Business,* 28 August 2006 (trade journal-online). Roberts supports the belief that the aftermath of Betsy saw community members helping each other with no expectation of government aid. After Katrina, however, the expectation was high that the federal government should aid in rebuilding. Landphair, "The Forgotten People of New Orleans." After Hurricane Betsy, the U.S. Army Corps of Engineers revamped its thinking regarding Mississippi River flooding. Instead of only focusing on levees to protect people from flooding, the Corps began building floodwalls to mitigate the flooding from storm surges. The engineering community also recognized the problem, with sinking levees contributing to the threat of storm surges and the need to monitor levee height. See Deon Roberts, "Hurricane Betsy Prompts Levee Upgrades" *New Orleans CityBusiness,* 4 August 2006, p. 1.

9 In approaching the topic of Katrina, it is easy to be overwhelmed by the volume of print and visual sources. The literature on Katrina includes scholars exploring the disaster from various perspectives, ranging from the incompetent and indifferent response of the federal government to the ecological realities that made flooding in the Ninth Ward inevitable to climate change, to name a few. One of the best recent books to be published, which also won the prestigious Bancroft Prize, is by Andy Horowitz, *Katrina: A History, 1915–2015* (Cambridge: Harvard University Press, 2020), in which he examined Katrina from multiple perspectives, including the inadequacy of past and present policies that not only contributed to Katrina but also impeded the recovery. But Horowitz also succeeds in revealing the individual experiences of so many whose lives were devastated by this man-made disaster. Another impressive text that touches upon several of the topics in this text, in addition to Katrina, is Kimberly Ruffin, *Black on Earth: African American Ecoliterary Traditions* (Athens: University of Georgia Press, 2010). For a good general history of Katrina, see Gary Rivlin, *Katrina: After the Flood* (New York: Simon & Schuster, 2015).

10 For statistics regarding New Orleans during the time of Katrina, including demographic data on the Ninth Ward such as poverty rates, home ownership, and the number of residents who owned cars, see Michael Eric Dyson, *Come Hell or High Water: Hurricane Katrina and the Color of Disaster* (New York: Basic Books, 2005; rpt. 2007). Dyson's statistics support his thesis on how race played a role in the treatment of those most affected by Katrina and aptly demonstrate the level of poverty that New Orleans's residents were experiencing. In Douglas Brinkley's account, he focused on Nagin's lack of leadership. See Douglas Brinkley, *The Great Deluge: Hurricane Katrina, New Orleans, and the Mississippi Gulf Coast* (New York: William Morrow, 2006);

Los Angeles Times, 30 August 2010 accessed at https://www.latimes.com/archives/la-xpm-2010-aug-30-la-na-0830-katrina-superdome-20100830-story.html. In the Paul Harris interview, he revealed how as a tourist, he and other tourists received different treatment than the others. Harris and an estimated 100 fellow tourists were escorted out of the Center after three days to the Hyatt Regency where the mayor had also evacuated. When Harris and the others were led out of the Center, they were advised not to smile or let out that they were being relocated to another site. Rochelle Smith Interview in *Overcoming Katrina*, p. 124.

11 For a discussion of New Orleans' increased vulnerability to hurricanes, see James F. Barnett, Jr., *Beyond Control: The Mississippi River's New Channel to the Gulf of Mexico* (Jackson: University Press of Mississippi, 2017); Landphair, "The Forgotten People of New Orleans"; Elizabeth Fussell, "The Long-Term Recovery of New Orleans' Population after Hurricane Katrina," *American Behavioral Scientist*, Vol. 59, No. 10 (September 2015):1231–1245 accessed at https://journals.sagepub.com/doi/abs/10.1177/0002764215591181; Ari Kelman, "In the Shadow of Disaster" *The Nation,* 15 December 2005 accessed at https://www.thenation.com/article/archive/shadow-disaster/; Rachel Nuwer, "Hurricane Katrina and Rita Caused at Least 117 Uncounted Deaths of Stillborn Babies" *Smithsonian Magazine*, 9 May 2014 accessed at https://www.smithsonianmag.com/smart-news/hurricanes-katrina-and-ritas-death-tolls-increase-when-stillbirths-are-taken-account-180951388/; Ann Fabian, "Seeing Katrina's Dead" *Katrina's Imprint: Race and Vulnerability in America*, ed. Keith Wailoo, et al. (New Brunswick: Rutgers University Press, 2010):59–68.

12 The idea that Katrina and its consequences were not new experiences for African Americans was voiced by Clement Alexander Price in his essay, "Historicizing Katrina." In Price's words, "Indeed, although Hurricane Katrina is being exceptionalized as the nation's worst, the social aftermath of the storm is hardly without precedent." *After the Storm: Black Intellectuals Explore the Meaning of Hurricane* Katrina, ed. David Dante Troutt (New York: The New Press, 2007):72. Still another commentator noted, "In New Orleans, the higher, safer ground has always been occupied by richer, white folk while the lower, more dangerous ground, has always been the province of poorer people." See Michael Eric Dyson, *After the Storm*, Part Three, (2007): Another reference to African American memories of the Red Cross in the 1920s and 1930s can be found in Susan Cutter, et al., *Hurricane Katrina and the Forgotten Coast of Mississippi* (New York: Cambridge University Press, 2014):982. For a discussion of the vulnerability of neighborhoods such as the Ninth Ward, see Landphair, "The Forgotten People of New Orleans"; Craig E. Colten, "Vulnerability and Place: Flat Land and Uneven Rise in New Orleans" (December 2006):731–734.

13 Lil' Wayne (2008) "Tie My Hands" accessed at https://www.azlyrics.com/lyrics/lilwayne/tiemyhands.html

14 K-Otix (2005) "George Bush Doesn't Care About Black People" accessed at https://genius.com/K-otix-george-bush-doesnt-care-about-black-people-lyrics

15 Courtney George, "From Bounce to the Mainstream: Hip Hop Representations of Post-Katrina New Orleans in Music, Film and Television" *European Journal of American Culture*, Vol. 35, No. 1 (2016):17–32; Jason (2011) "Hurricane Katrina Damage and Rap Song" accessed at https://www.google.com/search?q=Hurricane+Katrina+and+rap+song+on+youtube&rlz=1C1SQJL_enUS783US783&oq=Hurricane+Katrina+and+rap+song+on+youtube&aqs=chrome.69i57j69i60.8554j0j15&sourceid=chrome&ie=UTF-8

16 To learn more about Claire Crewe and her poetry, visit her website at https://www.clairecarew.com/index.php?vw=1458&vh=794&v0=&v1=128&v1b=0&v2=can&v3=0&v25=965

17 Yusef Komunyakaa, who was raised in Louisiana, is the recipient of several awards for his poetry. To learn more about the inspiration behind his poem, "Requiem" see https://poetrysociety.org/features/remembering-katrina/yusef-komunyakaa-2

18 For an in-depth view of Ted Ellis' art, visit his website at https://tedellisart.com/. More information on the Lower Ninth Ward Living Museum can be found at their website http://www.l9livingmuseum.org/

19 Willie Pitford Interview, *After Katrina*, p. 110; Keith Elder, et al., ""African Americans' Decisions Not to Evacuate New Orleans Before Hurricane Katrina: A Qualitative Study" *American Journal of Public Health*, Vol. 97, No. suppl. 1 (April 2007) accessed at https://ajph.aphapublications.org/doi/10.2105/AJPH.2006.100867; Rebeca Antoine, ed., *Voices Rising: Stories from the Katrina Narrative Project* (New Orleans: University of New Orleans Press, 2014):219. Another source for firsthand Katrina accounts is the Hurricane Digital Memory site at http://hurricanearchive.org/. For further discussion regarding the choice of some New Orleanians to remain in New Orleans and not evacuate, see Anja Nadine Klopfer, "Choosing to Stay: Hurricane Katrina Narratives and the History of Claiming Place-Knowledge in New Orleans" *Journal of Urban History*, Vol. 43, No. 1 (January 2017):115–139. In her article, she cites earlier accounts by New Orleans' Creole population regarding the river and their opposition to additional levees. The reasoning given behind the opposition signified the knowledge the Creole population possessed about the river and its behavior. While their "local knowledge"—and support for more outlets to absorb the floodwaters—clashed with the engineers and technocrats who advocated additional levees to prevent flooding, their relationship with the river complemented that of African Americans who decided to not evacuate.

20 Leonard Smith Interview, *After Katrina*, pp. 30–34.

21 Pete Stevenson Interview; Parnell Herbert Interview, *After Katrina*, pp. 37, 48; For further testimony regarding the belief in weakened levees and the suspicion that they were breached, see Hurricane Digital Memory site and Antoine, *Voices Rising.*

22 *The Times Picayune*, 28 August 2015; Denise Roubion-Johnson Interview, *After Katrina*, pp. 72–75. For further discussion regarding the term "refugee," see Fred Arthur Bonner II, "God's Gon' Trouble the Water: An African American Academic's Retrospective on Hurricane Katrina" *The Journal of Negro Education*, Vol. 75, No. 4 (Fall 2006):577.

23 Courtney Gisriel, "Survivor Stories; Family Reflects on how Hurricane Katrina Brought Them Closer Together" *TODAY,* 27 September 2018 accessed at https://www.today.com/news/survivor-stories-family-reflects-how-hurricane-katrina-brought-them-closer-t137527

24 "On This Block, Some Traditions Refuse to Die" *Washington Post,* 28 August 2015, accessed at https://www.washingtonpost.com/sf/national/2015/08/28/in-the-lower-ninth-some-traditions-refuse-to-die/; Reverend Sanders Interview, 24 January 2006 as quoted in Jackson, "Declaration of Taking Twice," pp. 778–779; Emily Chamlee-Wright and Virgil Storr, ""There's No Place Like New Orleans": Sense of Place and Community Recovery in the Ninth Ward after Hurricane Katrina" *Journal of Urban Affairs*, Vol. 31, No. 5 (12 January 2009):615–634.

25 *Washington Post*, 28 August 2015; Chamlee-Wright and Storr, "There's No Place Like New Orleans," pp. 615–634.

26 Fussell, "The Long-Term Recovery of New Orleans' Population after Katrina"; *Washington Post*, 23 August 2015; Horowitz, *Katrina: A History*, p. 195; *Times-Picayune, NO Advocate,* 8/29/20 accesses at nola.com.

27 Keith Ferdinand Interview in *After Katrina*, p. 99; Richard Mizelle, Jr., "Second-Lining the Jazz City: Jazz Funerals, Katrina, and the Reemergence of New Orleans" in Keith Wailoo, ed., et al., *Katrina's Imprint: Race and Vulnerability in America* (New Brunswick: Rutgers University Press, 2014):73. Another study examining the impact of Katrina upon displaced families can be found in Lori Peek, et al., "Disaster Hits Home: A Model of Displaced Family Adjustment After Hurricane Katrina" *Journal of Family Issues*, Vol. 32, No. 10 (2011):1371–1396.

28 Stéphane Tonnelat & translated by Eric Rosencrantz, "Making Sustainability Public: The Bayou Observation Deck in the Lower 9th Ward of New Orleans" *Metropolitics*, 20 June 2011 accessed at http://www.metropolitiques.eu/Making-sustainability-public-The.html

29 *New York Times,* 31 August 2015.

30 Cindy Ermus, ed., *Environmental Disaster in the Gulf South: Two Centuries of Catastrophe, Risk, and Resilience* (Baton Rouge: Louisiana State University Press, 2018); Jason David Rivera and DeMond Shondell Miller, "Continually Neglected: Situating Natural Disasters in the African American Experience" *Journal of Black Studies*, Vol. 37, No. 4 (March 2007):502–522; Collectif Argos, *Climate Refugees* (Cambridge: MIT Press, 2010):170; Keith Ferdinand, *Overcoming Katrina*, p. 92.

EPILOGUE

FIGURE E.1 Norco, Louisiana Refineries.
Credit: Spiderscanabis.

By the late 1900s and the early twenty-first century, while New Orleans and the surrounding coastal communities experienced devastating hurricanes followed by equally harmful storm surges, another part of the river corridor extending from Baton Rouge to New Orleans was undergoing significant industrial growth (Figure E.1). For an 85-mile stretch, in the area once known as "Plantation Country"—now referenced as "Cancer Alley," and more recently "Death Alley"—petrochemical industries began buying property in places such as the

DOI: 10.4324/9781315617077-7

Old Diamond neighborhood, part of Norco, a town approximately 25 miles west of New Orleans, and once home to sugarcane plantations. The new industry grew quickly, with 126 plants on the Lower Mississippi River in 1962, expanding to 196 in 2002. The riverfront sites were attractive to industry for several reasons, ranging from access to shipping lanes to protection from hurricanes. These petrochemical plants, however, threatened the very existence of numerous African American communities, such as Old Diamond, many with residents who had descended from enslaved people who had lived and worked on these lands during the antebellum era. Companies such as Royal Dutch Shell and DuPont bought the properties, creating fenceline communities where the residents were subject to polluted air and water, causing severe health problems, including an increase in cancer cases and respiratory illnesses. In response to their presence, these tight-knit communities protested the degrading of their environment and the threat to a quality of life that had been sustainable for generations. Their protests echoed a resistance dating back to the maroons and more recently, seen in the return of so many to rebuild New Orleans, reaffirming the resilience that is an integral theme of the African American history of the Mississippi River. Despite protests, however, the petrochemical industry has remained in the Lower Mississippi Valley, ushering in a new chapter with the Mississippi River.[1]

For African Americans who owned homes in the riverside communities and often worked in the industrial plants lining the waterway, their history with the river reflected international concerns with the environment with its precious waterways and the health of nearby residents. "Cancer Alley," though localized to the Lower Mississippi River Valley, represented the intersection of highly toxic industries—polluting the air and water—with their natural surroundings. For those living in the area, a host of health issues resulted, including higher rates of cancer, kidney and respiratory ailments, to name a few, products of the industrialized corridor and a system that rewards profit over sustainability. In addition to health consequences, the industrial corridor was also the site of a high number of devastating explosions. One report claimed that by the late twentieth century, "the Baton Rouge to New Orleans industrial corridor stood out as one of the most prominent zones of chemical plant explosions." Yet, in many ways, despite the international overtones, the differences from past intersections with the river were slight in that the treatment of the river and its environs always paralleled the treatment of African Americans whether enslaved or free. Just as the river was utilized to support commercial shipping culminating in a cotton trade that amassed millions for antebellum planters, a slave regime was instituted, extracting the labor of millions of African Americans. After slavery, the same river would be manipulated and controlled to foster economic growth while African Americans were confined to back-breaking labor building levees or sharecropping the bottomlands. By the twenty-first century, the disregard for the ecological health of the Lower Mississippi River Valley paralleled the petro-chemical industry's indifference to the health of African Americans living there. However, by the 2000s, protests against those contaminating the valley were louder and more articulate than earlier resistance efforts, such as those of the villagers living in Fazendeville or the 1927 Flood survivors, yet building upon a culture of resistance that began in the 1700s with a maroon culture that defied the slave regime. Nonprofit governmental organizations, such as RISE St. James, under the leadership of long-time residents of St. James Parish, such as Sharon Lavigne, are calling attention to the abuse of the petrochemical factories. Adding to these voices are international organizations, such as the United Nations, demanding an end to the environmental racism found in "Cancer Alley."[2]

As there are countless actors resisting the actions of the petrochemical industries in the lower Mississippi River corridor, extracting only a few stories is challenging. But two activists, in particular, received international recognition for their environmental activism. Both Margie Richard of the Old Diamond neighborhood and Sharon Lavigne of Saint James Parish received the Goldman Environmental Prize. Richard was awarded the prestigious prize in 2004 whereas Lavigne was selected in 2021. Compared with the environmental equivalent to the Nobel Prize, the Goldman Environmental Prize

> honors grassroots environmental heroes from roughly the world's six inhabited continental regions: Africa, Asia, Europe, Islands & Island Nations, North America, and South & Central America. The Prize recognizes individuals for sustained and significant efforts to protect and enhance the natural environment, often at great personal risk.

For communities living along "Cancer Alley," the prize also meant increased awareness of the abuses of the giant petrochemical industry.[3]

With a long-time connection to Diamond, Margie Richard was the fourth generation to live in the neighborhood. Her roots in the area, however, date back even further, as she is a descendent of those who participated in the Deslondes Revolt in 1811, the largest slave revolt to occur in the United States. As discussed in Chapter 2, the revolt was suppressed violently with those convicted, executed and their heads severed and displayed on pikes lining the river. Growing up in Diamond in the 1940s, Richard recalled running up and down the levee as well as the nearby swamp. She also remembered her grandfather and other African American residents in the neighborhood growing vegetables in their gardens, attaining a level of independence through the rich riverine environment. For many African Americans, such as Richard, their ancestors had once been enslaved on the Trepagnier Plantation, with an economy dominated by the production of sugarcane. After the Civil War, many of the formerly enslaved stayed in the area and like Margie Richard's family were living in Diamond when Shell Oil arrived in 1916. Initially, Diamond residents were pleased when Shell constructed a nearby refinery, expecting employment with the company. The few jobs available, however, were unskilled. Thus, the arrival of Shell Oil brought little to no benefits for Diamond inhabitants.[4]

For Royal Dutch Shell, however, the year 1916 was only the beginning of their presence in the Lower Mississippi River Valley. By the 1950s, the corporation was purchasing land from long-time residents as the Shell Chemical Plant was being built. Now for those who remained, they lived in what were known as "fenceline" communities, areas bordering the newly built chemical plant and refinery. Shell, similar to the growing number of petrochemical plants ensconced on the lower Mississippi River corridor, was attracted to the area for several reasons. First, the Lower Mississippi River, known for its depth, could accommodate oceangoing vessels upstream and offered a safe haven inland, free from the hurricanes that landed on the Gulf Coast. Second, in the 1950s and up until the late twentieth century, company officials—supported by experts—held that the Mississippi River's volume acted to dilute the toxic wastes spewed by the plants. Third, Louisiana State provided a welcoming environment for corporate interests, with a generous tax structure, which was already established earlier with oil and gas corporations. Given these advantages, communities such as Diamond became, in the words of environmental justice scholars, "sacrifice zones." For the people remaining in Diamond, by the mid-1970s their property diminished in value, making relocation impossible. But Diamond became a sacrifice zone for other reasons as well. For those whose homes butted up against the chemical plant and refinery, the residents were experiencing a host of medical issues caused by the plant's toxic chemical releases. Not only did the air smell but also Diamond residents complained of headaches, watering eyes, sinus problems, asthma, and higher than normal cancer rates and birth defects.[5]

As a long-time Diamond resident, Margie Richard witnessed what was happening to her neighbors as well as at home. She lost a sister, Naomi, at the young age of 43, a victim of sarcoidosis, a bacterial infection. Although sarcoidosis is rare, with an infection rate of one in a thousand, Richard knew three other neighbors who contracted the infection. But the occurrence of the life-threatening illnesses was not the catalyst for Richard's activism. Instead, what finally prompted her to protest the actions of Shell was an incident in 1973 when a pipeline leak resulted in an explosion at the home of a Diamond resident, resulting in two painful deaths: the homeowner and the teenage boy who had been mowing her lawn. Another explosion occurred in 1988, killing seven workers, which convinced Richard that residents should not be in such close proximity to the Shell plant. At this point, Richard began to rally neighbors, forming the Concerned Citizens of NORCO, which would challenge the actions of Shell and succeed in redressing some of its worst offenses. The goals of the Concerned Citizens of NORCO included the reduction of toxic emissions from Shell Chemical, financial support from Shell for relocation for Diamond residents, and a community development fund. Through her tireless advocacy, Richard succeeded in these goals and by the time Richard received the Goldman Prize in 2004, Shell had reduced emissions by 30%, provided relocation funds averaging $80,000 per homeowner, and contributed $5 million for a community development fund. Since her initial work with Concerned Citizens, Richard has become an international celebrity in environmental justice circles, implementing the work of other environmental organizations, such as the Bucket Brigade. (The Bucket Brigade trains local citizens how to test air quality, among other activities.) Richard's story is testimony to a relationship with the river and its surroundings that had once offered sustenance and recreation, a linkage that prompted resistance when jeopardized. In this instance though, despite the efforts of Richard and Concerned Citizens, by 2002 the entire community of Diamond had been purchased by Shell. While the residents could claim victory in that they secured a buyout from Shell, allowing them to leave what had become a toxic chemical environment, their success also meant the abandonment of a community with a history dating back three centuries.[6]

Another compelling story of someone who protested the actions of the petrochemical industry is that of Sharon Lavigne, whose family history with the riverside community in the Fifth District of St. James Parish, a predominantly African American population, extends back to the antebellum period. Lavigne, also a Goldman Prize recipient in 2020, remembered a time when her community could boast of clean air and fertile land for growing fruits and vegetables. In her words, "The first time I went to the Mississippi River I was six. All I could feel was fresh air flowing through giant grasses filled with blackberry vines and white spider lilies. We had everything you could imagine." She described a rich environment where her grandfather caught fish and shrimp from the river. But now, Lavigne lamented ". . . the land and everything that grows on it is poison." She went on, "We are boxed in from all sides by petrochemical plants, tank farms, and noisy railroad trucks. We are sick. So many people are dying of cancer, upper respiratory diseases, asthmatic conditions and on and on." According to the U.S. Environmental Protection Agency (EPA), St. James Parish had "one of the highest concentrations of toxic chemicals in the country with cancer rates 50 times higher than the national average." In addition to dealing with the disease itself, Lavigne also referenced the medical bills that were incurred as a result. She suggested that the title "death row" would be more appropriate than "Cancer Alley," as a designation for the expanse between Baton Rouge and New Orleans; with its profitable sites for an estimated 150 petrochemical plants.[7]

The catalyst, however, that prompted Lavigne's activism came when the Chinese chemical firm, Wanhua, proposed building a plastics manufacturing plant for a cost of $1.25 billion in November 2018. The plant would produce MDI, which, along with other toxins associated with this chemical, causes cancer and respiratory ailments. In response to the proposed plant, Lavigne decided to organize the grassroots organization, RISE St. James, with the goal of stopping construction. Through RISE St. James, Lavigne collaborated with other environmental and faith-based organizations. What began as a small meeting with only ten in attendance grew into a successful undertaking that ultimately stopped Wanhua, which withdrew its plans by September 2019. But Lavigne's community would still be threatened by another petrochemical corporation as the Taiwanese firm, Formosa Plastics, announced its intent in 2015 to construct a $9.4 billion plastics manufacturing plant. Again, Lavigne and her organization responded. One of the first complaints against Formosa's proposed development was their location on burial grounds with the remains of enslaved people from antebellum plantations that once dominated the area. RISE St. James claimed seven grave sites in the area, with only four on the land where Formosa planned to build. Formosa, however, only acknowledged two sites and said only one contained any graves. Lavigne and others in St. James Parish gained national and international attention, with the United Nations decrying the threat to the burial sites. In a statement by UN experts, they voiced concern over what they identified as "possible violations of cultural rights, as at least four ancestral burial grounds are at serious risk of being destroyed by the planned construction."[8]

In addition to their commitment to save burial sites, RISE St. James also persevered in its opposition to the plant based on the health of residents in the Fifth District of St. James Parish. Again, their work drew the attention of national and international audiences. In one article, published in 2021 by *The Atlantic*, the author cited historian Barbara Allen's observations about the presence of the petrochemical industries. Allen recognized the historical continuity for African Americans living along the Lower Mississippi River. The title of the article, "One Oppressive Economy Begets Another," was first voiced by Allen as she remarked, "The Great River Road was built on the bodies of enslaved Black people. The chemical corridor is responsible for the body burden of their descendants." On an international level, the United Nations not only condemned Formosa's proposal and its effect upon burial sites but also framed Formosa's actions as another glaring example of environmental racism. The international organization called for an end to any further industrialization of the industrial corridor, stating that the presence of the pollution-emitting chemical plants was a form of "environmental racism."[9]

Other political actors, such as Michael S. Regan, the EPA Administrator, under the Biden presidency, joined the growing number protesting the effects of industrial behemoths, such as Formosa, and their impact upon African American communities along the Lower Mississippi River. As a result, by fall 2021, the U.S. Army Corps of Engineers imposed a new, more rigorous review of the project, forestalling development for another 2 years. Lavigne, like Margie Richard, illustrated the persistence of long-time Mississippi River communities whose roots can be traced to the days of enslavement. For Lavigne and Richard, their childhoods and the livelihoods of their parents were enriched by the river's largesse. Corporations, such as Formosa and Royal Dutch Shell, threatened a way of life—and in the case of Old Diamond destroyed—that relied upon this largesse while also fouling and polluting the river itself. In this new chapter of the Mississippi River and its intersection with the lives of African Americans, the river as a source of revenue—a constant in the river's past—signaled the demise of sustainable African American communities.[10]

FIGURE E.2 View of Holy Rosary Cemetery, Taft, Louisiana with Petrochemical Plant in Background.
Credit: Tiago Fernandez Photography.

But St. James Parish and Diamond were not the only communities to be either threatened or destroyed by the presence of petrochemical plants in the Lower Mississippi River corridor (Figure E.2). Reports of increased cancer rates, respiratory ailments, birth defects, and a deteriorating physical environment were being reported in the late 1980s and 1990s in places such as St. John the Baptist Parish. In one firsthand account from parish resident Amos Favorite, he warned the residents of St. John that if they don't fight now, they will become another Geismar. "I had a garden out back there," he recalled, gesturing at the land behind his brick home.

> One night, there was a heavy fog out there, I thought it was a heavy fog. I woke up the next morning, and everything was shriveled up and dead. That is enough proof for me that these chemicals are killing us.

Downriver, Wilfred Greene had seen neighbors sicken and die. But even those who remained healthy faced the loss of a treasured lifestyle, as Greene lamented an earlier life before the arrival of the petrochemical industry,

> If I didn't have money and wanted to fix myself supper, I could just go out and shoot a rabbit or a bird, catch myself a fish. Now, I catch fish so oily and slimy I wouldn't think of eating it. My dog wouldn't eat it.

Greene lived by the levee on the land that his great-grandparents bought after Emancipation. His small brick home was paid for.

> It may not be as good as some people have, but to me, it's home . . . It's peace. I don't owe anybody. If I don't want to open the door in the morning, I don't have to. If I want to walk in the backyard and play with my dog, I can. If I want to plant butterbeans, I can.

Adding to Greene's sense of history was a nearby cemetery where his relatives were laid to rest, beneath cypress trees a century old.[11]

The pollution, however, by a robust petrochemical industry, aided in large part by a generous state tax structure persisted. More in-depth coverage of the abuses resulted and beginning in May 2019, *The Guardian* published a series of articles on Cancer Alley, beginning with an insider's perspective of the African American community of Reserve, also located in St. John the Baptist Parish. Unincorporated, Reserve was the home of many whose ancestors had been enslaved on what were once plantations lining the river. After the Civil War, many formerly enslaved people remained, with some able to eventually buy land and build homes. This was the case for Joseph James, Sr. who built his own home in the parish along the River Road, only to see his neighborhood threatened when DuPont constructed its plant nearby in 1968. DuPont's plant produced the chemical, chloroprene, which was used primarily in manufacturing neoprene. A key ingredient in wet suits, laptop sleeves, and even beer koozies, among other goods, the Louisiana plant is the only one in the United States that produces neoprene. But for the residents of St. John the Baptist Parish, the costs of producing neoprene are deadly as the U.S. EPA documented that "the five census tracts around the plant have the highest cancer rates in the country—more than seven hundred times the national average in one tract." Aware of the toll that the chemical plant incurred on their lives along with a keen sense of history, James' daughter, Mary Hampton observed, "When you think about it, nothing has ever really changed," she says. "First slavery, then sharecropping, now this. It's just a new way of doing it." She recognized the boom-and-bust cycles the area was subject to as an extractive economy dominated, commodifying the land and the people.[12]

Adding to Hampton's critique is Robert Taylor, another outspoken resident of the parish and founder of Concerned Citizens of St. John the Baptist Parish. Taylor, an 80-year-old tireless advocate against Denka, which bought the chloroprene chemical plant in 2015, recounted the effects of living nearby the complex. Having grown up in St. John the Baptist Parish in a house built by his father, Taylor lost his mother to bone cancer, a brother to lung cancer, a sister to cervical cancer, a nephew to lung cancer, and both neighbors on one side of his home to cancer, and one neighbor on the other side. When he learned about the EPA data documenting the high cancer rates in St. John the Baptist Parish,

Taylor, through the Concerned Citizens organization, began asking Denka to reduce emissions. While Denka responded with a reduction, the reduction did not meet the 0.2 guideline based on the Integrated Risk Information System, an assessment program developed by the EPA. By May 2021, Taylor, determined to reclaim a livable, sustainable environment, petitioned the Inter-American Commission on Human Rights, an agency within the Organization of American States, "to investigate cancer risks, recommend air pollution limits and join the group's calls to shut down the plant." In Taylor's words, "We suffer and die because they decided to dump poison into our air. We live in the state's sacrifice zone."[13]

But in Taylor's critique of Denka, he also considered the role of the river in St. John Parish. Unlike Richard, who remembered a childhood playing on the levee with children running around the swamp, or Lavigne, who presented a Mississippi River environment, rich with vegetation, sustaining her community through fishing and gardening, Taylor considered the river's historical productive role. He reviewed how the area where he lived and his ancestors were once enslaved was fueled by the sugarcane industry. Taylor said that even after "supposed" emancipation, their lives were based around the sugarcane plantations. In his words, "The reason these plantations were so prosperous was the river. They had to distribute their products. It's the same with industry." He went on to reflect how most societies were built around rivers and then concluded that once the planters realized they could profit from the petrochemical industry, they "pulled up stakes." In Taylor's worldview, the river, in its historical role, was always one of the determinants—facilitating large-scale economies—for those African Americans living in St. John the Baptist Parish. And for those still living in the parish, Taylor contended that the actions of the petrochemical industry amounted to "genocide." In support of this statement, Taylor observed, "92% of the population that is impacted by the petrochemical industry is black in a state where we're only 32% of the population."

The accusation was expressed in other communities that lined the river. For example, Anne Rolfe, director of the Louisiana Bucket Brigade, a nonprofit organization dedicated to employing "grass roots action to hold the petro-chemical industry and government accountable for the true costs of pollution," contended that the Fifth District in St. James Parish was slated for elimination. In her words, within the proposal by Formosa to build, "there's also an unspoken companion plan to wipe out the Fifth District. They are very clearly eliminating the black community. I've been working on the river for 20 years and this is the most egregious I've ever seen." Still another critic of corporate actions is Emilie Townes, who, in even more damming language, compared the siting of toxic waste landfills in African American communities with "contemporary versions of lynching a whole population." In response to these threats to African American communities and the degradation of the Lower Mississippi Valley, the Bucket Brigade integrates a historical component in the organization. In their work preserving the land and communities, their staffing includes a genealogist who assists those "with ancestral ties to the parishes along the Mississippi River." Through the sharing of genealogical expertise, African American families can reclaim their history and preserve the sites where their families were buried. Also on the staff is a Reparative Justice Legal Fellow, whose work includes the preservation of the "region's rich cultural memory" among other jobs. The organization also sponsors bike rides in the river parishes, cultivating an "exploration of Black History and the environmental justice movement." The inclusion of these positions and activities acknowledges distinct African American perspectives of the river—as a sense of history with linkages to the river dating back centuries.[14]

In response to the work of grassroots organizations, such as RISE St. James, Concerned Citizens of St. John the Baptist Parish, and the Bucket Brigade, calling attention to these dire conditions and intent on holding corporations accountable, Michael Regan, the EPA administrator, visited communities in "Cancer Alley," in his Journey to Justice tour. (Regan also visited other areas in the South experiencing air and water pollution.). After his tour in November 2021, Regan promised to commit federal infrastructure funding to address the racial disparities found in communities burdened with air and water pollution. By January 2022, in fulfilling this promise, Regan announced that $600,000 in federal funds would be used to purchase air pollution monitoring equipment as well as begin unannounced inspections of chemical plants and refineries. He also recommended hiring additional inspectors for monitoring the air in Robert Taylor's St. John the Baptist Parish. Heartened upon hearing the news, Robert Taylor observed, "we are overwhelmed at the response we are getting from this administration." But Sharon Lavigne introduced a note of caution when upon hearing the administration's commitments, she remarked, "just the beginning of what needs to be done." Whether sustainable or not, this recent commitment signaled an improvement from previous administrations and federal policy, actions that probably would not have occurred without the activism of residents such as Lavigne and Richard.[15]

But African Americans were not the only casualties of industrial growth in the Lower Mississippi Valley. Adding to the devastation wrought by the petrochemical industry, the Lower Mississippi River Valley was experiencing the loss of its wetlands, again a victim of human actions. As discussed in Chapter Five, without the wetlands, which acted as a buffer, the destruction from hurricanes has worsened. Yet, the wetlands continue to shrink, with the oft-cited reference that for every 100 minutes, Louisiana loses wetlands equal to the size of a football field. Communities such as Isle de Jean Charles—homeland to the Isle de Jean Charles Biloxi-Chitimacha-Choctaw Tribe—are being subsumed by rising sea levels, with their residents facing relocation. This ecological disaster has long-standing causes as the engineering of rivers coupled with a missionary zeal to modernize became a gold standard in the United States. For the lower Mississippi River, the wetlands deteriorated as a result of levee-building, which began in the early eighteenth century, eventually robbing the Delta of the sediment that the river deposited annually. Added to this was the boom in oil production, as Louisiana offered an attractive tax structure for oil companies that resulted in dredging the wetlands and building a system of canals and pipelines that allowed for shipping lanes to the Gulf Coast. The construction of the canals, such as the Mississippi River Gulf Outlet and the Gulf Intercoastal Waterway, allowed saltwater to find its way into the wetlands, further undermining the environment. According to experts, the canals and pipelines accounted for 36% of the loss of wetlands. But climate change also contributed to the loss of wetlands as sea level rises combined with more severe hurricanes and the subsequent storm surges in recent years. Finally, as the coastline erodes and cities such as New Orleans continue to sink, other problems arise such as the aging drainage systems that are unable to keep in check increasingly heavy rains. Confronted with these challenges, scholars have posed the question of whether New Orleans and other vulnerable coastal communities should be protected. Yet the determination of neighborhoods to rebuild after Katrina, such as the Ninth Ward, undermines the viability and morality of these considerations.[16]

In framing the challenges introduced by the petrochemical industry, coincident with the construction of levees and the consequences of climate change is a growing body of scholarship on environmental justice. The subdiscipline—a union between environmental activists and scholars—began in Warren County, North Carolina in the 1980s when local activists argued against the siting of a toxic waste dump in their neighborhood, home to a majority African American population. Their opposition attracted national attention and soon scholars, such as Robert D. Bullard, were writing about the toxic waste proposal, presenting it within the context of environmental racism. In Bullard's words, the definition of environmental racism

> . . . refers to any policy, practice, or directive that differentially affects or disadvantages (whether intended or unintended) individuals, groups, or communities based on race or color. It also includes exclusionary and restrictive practices that limit participation by people of color in decision-making boards, commissions, and regulatory bodies. Environmental racism exists within local zoning boards as well as the U.S. Environmental Protection Agency.

Following Bullard were others who recognized the links among polluted landscapes and people of color and the economically disadvantaged in the United States. These observations have resulted in a growing body of environmental justice studies that have expanded into the arena of climate change. For those researching climate change, there is a growing awareness of the disproportionate effect of climate change upon people of color. Further for scholars of climate change, prompting what is now referenced as the Anthropocene Age, the displacement of local populations has emerged as one of the most alarming outcomes of global warming. The forced internal displacement of Katrina survivors testified to this disturbing outcome. Thus, as the literature evolved, environmental justice studies expanded to include more than polluted landscapes as scholars explored the effects of polluted air and water; urban green spaces; and the development of policies favoring some while disenfranchising others. New scholars contributed research, revealing in the words of Rob Nixon, "the slow violence" of today's corporate actions and how communities such as those in the Lower Mississippi Valley were suffering disproportionately as a result of corporate greed.[17]

Still another extension of environmental justice studies is the overlap of environmental justice theories with critical race theory (CRT) and its underlying premise that racism is integral to American institutions and society. Extending the arguments of CRT, environmental justice studies have evolved into a critical environmental justice perspective. Building upon this perspective, environmental justice scholars, such as David Pellow, contend that there is a natural union between their research and the theoretical underpinnings of Black Lives Matter (BLM). Proponents of BLM contend that the violence against African Americans is state sanctioned, with Black lives "systematically and intentionally targeted for demise." Pellow advocates an understanding of "environmental racism as an extension of those state-sanctioned practices—in other words a form of authoritarian control over bodies, space, and knowledge systems." For the history of African Americans and the Mississippi River, the connection is well founded. Ultimately, environmental justice scholarship highlights the inequalities of "Cancer Alley" to a larger audience. But what does this mean for African Americans living along the Mississippi River? For those whose homes line the industrial corridor, they suffer the fate of policies that alternate between a progressive agenda, promising a cleaner environment, such as that offered by the Biden administration, and policies favoring large corporations, dismissing the claims of community members. For example, under the Trump administration from 2016 to 2020, corporations were allowed leniency in their monitoring due to the COVID pandemic. Without consistent policies, dictating limits for petrochemical industries, the health and welfare of the communities will always be jeopardized. Still, articulating their circumstances within the theoretical framework of the CRT adds another facet to the work undertaken by groups such as the Bucket Brigade or the Concerned Citizens of St. John Parish. The arguments employed by environmental justice scholars and expressed by international organizations, such as the United Nations, add a legitimacy to claims made by those whose lives have been upended by the petrochemical industry.[18]

In conclusion, by looking at the Mississippi River through the lens of African American history—a history shaped by centuries of enslavement, followed by a Jim Crow system that resulted in incarceration for some, sharecropping for others on the most vulnerable, flood-prone lands, back-breaking work on levees or "rousting" on steamships, and decades of lynching only to be succeeded by the actions of multinational corporations polluting long-existing riverfront communities—a different narrative of the river emerged. In its multiple roles, this Mississippi River shaped a history that has often been omitted from the dominant narrative. A bastion of freedom, the river also represented bondage and unchecked cruelty to millions of enslaved African Americans. Yet, more than a succession of binary experiences, African Americans' connection to the river produced a culture triumphant in its resilience, employing the river and its environs for sustenance and refuge, supporting communities such as Fazendeville, the Ninth Ward, and Old Diamond. This relationship with the river also produced heroes, leaders, and activists. Beginning with the legendary maroon, Jean San Malo, defying the slaveholding regime for more than 20 years to the bravery of those leading the Deslondes Revolt, heroes were born and commemorated through song and oral histories. Less well-known heroes were those who either successfully escaped enslavement using the river as an escape route or died trying. Contemporary leaders and activists include those such as Robert Taylor or Margie Richard, people committed to safeguarding lives in their riverfront communities by bringing an awareness to a culture fueled by profit, irrespective of the toll upon individuals and their communities. As so many African Americans' fortunes were tied to the river, this history deserves retelling. The river's presence also contributed to the literary accomplishments of writers, such as Richard Wright, who, through stories about the 1927 Flood, understood the river's role in shaping African Americans' lives. Yet the river also informed another subculture, one enriched by the music of groups as varied as the roustabouts to twentieth-century blues singers. Linkages to the river were also prompted by the need for sustenance and refuge, informing personal histories recalling an ecosystem that supplemented the meager diets of enslaved people or the nearby swamplands offering a respite from the horrors of slavery. By looking at the river through the lens of African Americans' experiences with the river, another history of the United States and global cultures surfaces.

Unfortunately, however, this history also supports the claims of movements, such as BLM, with its emphasis on state-sanctioned violence against African Americans as the river's legacy as a site for violence and trauma endures. But the history also uncovered centuries of interacting with the river and how the river not only sustained populations but was also a source of recreation, pleasure. Finally, just as the river served as an icon, a symbol of American exceptionalism in mainstream U.S. history, for African Americans, such as Henry Bibb, the river not only evoked a vision of freedom but also possessed its own beauty for all to enjoy. Ultimately, this Mississippi River, though not as uplifting as some river histories, adds a more inclusive, mature understanding of the past—demonstrating the multilayered roles that rivers play in our lives—and adds one more piece to the "patternless puzzle" that Hodding Carter recognized in earlier Mississippi River histories.

Notes

1 Craig E. Colten, "The Rusting of the Chemical Corridor" *Technology and Culture*, Vol. 47, No. 1 (January 2006):95. In Colten's article, he cites 196 refineries and chemical plants along the Lower Mississippi River by 2002. For figures that specifically document the number along "Cancer Alley," the estimated number is 150. There are

numerous scholarly overviews of development along the Lower Mississippi Valley; two frequently cited ones are Barbara Allen, Barbara L. Allen, *Uneasy Alchemy: Citizens and Experts in Louisiana's Chemical Corridor Disputes* (Cambridge: MIT, 2003) and Dorceta E. Taylor, *Toxic Communities: Environmental Racism, Industrial Pollution, and Residential Mobility* (New York: New York University Press, 2014).

2 Craig E. Colten, "An Incomplete Solution: Oil and Water in Louisiana" *Journal of American History*, Vol. 99, No. 1 (June 2012):94. The presence of so many petrochemical companies on the Lower Mississippi River prompted the formation of numerous local groups, intent on reclaiming their communities. One of the effective attempts has been the establishment of the Coalition Against Death Valley in March 2019. Its agenda included "a process of non-violent protests to pressure industrial giants and governments to stop the ongoing poisoning of majority-black communities in Louisiana's "Cancer Alley."" Members of the coalition include RISE St. James, Concerned Citizens of St. John, 350 New Orleans, Louisiana Bucket Brigade, Justice Beyond, Healthy Gulf, Earth Works, Center for Biological Diversity, Extinction Rebellion, and The Climate Reality Project. See https://www.enddeathalley.org

3 For more information regarding the Goldman Environmental Prize and the organization's work, visit their website at https://www.goldmanprize.org

4 For a biographical overview of Richard's life, see Margie Richard, Goldman Prize Foundation accessed at https://www.goldmanprize.org/recipient/margie-richard/. For an overview of African American attitudes about employment with the petrochemical industries, along with supporting data see Gregory R. Berry, "Organizing Against Multinational Corporate Power in Cancer Alley: The Activist Community as Primary Stakeholder" *Organization and Environment*, Vol. 16, No. 1 (March 2003):3–33.

5 The history of the Old Diamond neighborhood can be found in Steve Lerner, *Diamond: A Struggle for Environmental Justice in Louisiana's Chemical Corridor* (Cambridge: MIT Press, 2006). A discussion of state incentives for petrochemical interests can be found in Colten, "The Rusting of the Chemical Corridor." Concerns regarding pollution of the river were alleviated as "rivers with a high discharge rate were more resistant to pollution." Allen, *Uneasy Alchemy*, p. 12.

6 Lerner, Diamond, pp. 32–35, 55–65, 245–260. For an overview of the Bucket Brigrade and its work, visit https://labucketbrigade.org

7 For a biographical overview of Sharon Lavigne see the Goldman Prize Foundation, accessed at https://www.goldmanprize.org; "The World We Need: Sharon Lavigne's Story of RISE St. James," YouTube accessed at www.ncronline.org For another discussion of the concerns of RISE St. James, visit https://earthjustice.org/features/cancer-alley-rises-up

8 "This is environmental racism": Activists call on Biden to stop new plastics plants in "Cancer Alley" *The Guardian*, 17 May 2021; "Environmental Racism in Louisiana's 'Cancer Alley', Must End" *UN News*, 2 March 2021, accessed at https://news.un.org/en/story/2021/03/1086172. In the same article, the practices of petrochemical plants were condemned further as the writer stated, "The African American descendants of the enslaved people who once worked the land are today the primary victims of deadly environmental pollution that these petrochemical plants in their neighborhoods have caused." See *UN News*, 2 March 2021.

9 Anya Groner, "One Oppressive Economy Begets Another" *The Atlantic*, 7 May 2021 accessed at https://www.theatlantic.com/culture/archive/2021/05/louisiana-chemical-plants-thriving-off-slavery/618769/; *UN News*, Environmental Racism in Louisiana's 'Cancer Alley.'

10 *Washington Post*, 28 November 2021; "Army Corps Orders Environmental Review of Proposed Formosa Plastics Plant in Louisiana 'Cancer Alley' *resilience*," 19 August 2021 accessed at https://www.resilience.org/stories/2021-08-19/army-corps-orders-environmental-review-of-proposed-formosa-plastics-plant-in-louisianas-cancer-alley/

11 Ginny Carroll, "When Pollution Hits Home" *National Wildlife Federation*, 1 August 1991 accessed at https://www.nwf.org/Magazines/National-Wildlife/1991/When-Pollution-Hits-Home. For an overview of the demise of other African American communities as a result of the petrochemical industry in the industrial corridor of the Lower Mississippi River, see Barbara L. Allen, "Cradle of a Revolution? The Industrial Transformation of Louisiana's Lower Mississippi River" *Technology and Culture*, Vol. 47, No. 1 (January 2006):112–119.

12 "Almost Every Household has Someone That Died from Cancer" *The Guardian*, 6 May 2019; National Public Radio, "After Decades of Air Pollution, A Louisiana Town Rebels Against a Chemical Giant" 6 March 2018 accessed at https://www.npr.org/sections/health-shots/2018/03/06/583973428/after-decades-of-air-pollution-a-louisiana-town-rebels-against-a-chemical-giant

13 "St. John Parish Group Files International Human Rights Appeal Over Denka Area Pollution" accessed at https://www.nola.com/news/environment/article_685c8b2e-b72d-11eb-bae1-333baf276d13.html

14 University Network for Human Rights, "What is Cancer Alley?" (2019) accessed at https://www.humanrightsnetwork.org/waiting-to-die/background; Poor People's Campaign, "We Have Been Designated a Sacrifice Zone" (n.d.) accessed at https://www.poorpeoplescampaign.org/we-cry-power/robert-taylor/; Earth Justice, "Cancer Alley Rises Up" 14 February 2020 accessed at https://earthjustice.org/features/cancer-alley-rises-up; Emilie Townes as quoted in James H. Cone, *Risks of Faith: The Emergence of a Black Theology of Liberation, 1968–1998* (Boston: Beacon Press, 1999):139. For an overview of the mission of the Bucket Brigade and their advocacy work, see https://labucketbrigade.org/about-us/

15 "Cancer Has Decimated Our Community" *Washington Post*, 28 November 2021; "E.P.A. Chief Vows to 'Do Better' to Protect Poor Communities" *New York Times*, 26 January 2022; EcoWatch, "EPA to Address Air and Water Pollution in 'Cancer Alley'" 27 January 2022 accessed at https://www.ecowatch.com/epa-cancer-alley-pollution.html; 4WWL, "EPA Acts to Curb Air, Water Pollution in Poor Communities" 27 January 2022 accessed at https://www.wwltv.com/article/news/local/epa-acts-to-curb-air-pollution-in-poor-communities/289-90b45661-c92d-43db-af6f-c37c6d0f568b.

16 For a layman's introduction to the loss of wetlands in the Mississippi Delta, see Nathaniel Rich, "The Most Ambitious Environmental Lawsuit Ever" *New York Times Magazine*, 14 October 2014; John Wennersten and Denise Robbins, *Rising Tides: Climate Refugees in the Twenty-First Century* (Bloomington: Indiana University Press, 2017):87–92; "The Drowning Coast" *New York Times*, 25 February 2018. To learn more about the Isle de Jean Charles, Wennersten and Robbins, *Rising Tides* pp. 86–87. For insight into how the Mississippi River has been viewed by policy makers and the effect on wetlands, see Chavid Lavin and Chris Russell, "The Buoyancy of Failure: Battling Nature in New Orleans" *Space and Culture*, Vol. 9, No. 1 (February 2006):48–51. For a discussion on climate change and its disproportionate effect upon people of color, see W. Malcolm Byrnes, "Climate Justice, Hurricane Katrina, and African American Environmentalism" *Journal of African American Studies*, Vol. 18, No. 3 (September 2014):305–314.

17 Robert D. Bullard, "The Threat of Environmental Racism" *Natural Resources & Environment*, Vol. 7, No. 3 (Winter 1993):23–26, 55–56; Rob Nixon, *Slow Violence and the Environmentalism of the Poor* (Cambridge: Harvard University Press, 2011). In his words, "By slow violence, I mean a violence that occurs gradually and out of sight, a violence of delayed destruction that is dispersed across time and space, an attritional violence that is typically not viewed as violence at all." p. 2. Nixon also makes an excellent point, on how you can be displaced without moving, "dislocated lives in place," p. 20. A few classics in environmental justice scholarship include Robert D. Bullard, *Dumping in Dixie: Race, Class, and Environmental Quality* (Oxford: Westview

Press, 1990); Robert Bullard and Beverly Wright, *Race, Place, and Environmental Justice after Hurricane Katrina: Struggles to Reclaim, Rebuild, and Revitalize New Orleans and the Gulf Coast* (Boulder: Westview Press, 2009). For an overview of how the environmental justice movement grew, see Donna Houston, "Crisis is Where We Live: Environmental Justice for the Anthropocene" *Globalizations*, Vol. 10, No. 3 (2013):439–450; Luke W. Cole and Sheila R. Foster, *From the Ground Up: Environmental Racism and the Rise of the Environmental Justice Movement* (New York: New York University Press, 2001); J. Chakraborty, et al., "Environmental Justice Research: Contemporary Issues and Emerging Topics" *International Journal of Environmental Research and Public Health* (1 November 2016) accessed at https://doi.org/10.3390/ijerph13111072. In discussing climate change, consult the work of the United Nations Intergovernmental Panel on Climate Change, which provides a good introduction, accessed at https://www.ipcc.ch/reports/.

18 David Pellow, "Toward a Critical Environmental Justice Studies: Black Lives Matter as an Environmental Justice Challenge" *Du Bois Review—Social Science Research on Race*, Vol. 13, No. 2 (1 October 2016):1, 13.

BIBLIOGRAPHY

Archives/Special Collections/Museums

Delta State University Library, Cleveland, Mississippi
Herman T. Pott National Inland Waterways Library, University of Missouri-St. Louis
Historic New Orleans Collection—Williams Research Center
Lower Mississippi Valley Collections, Louisiana State University Libraries
National Archives, Kansas City Branch
New Orleans Public Library
State Historical Society of Missouri
Tennessee State Archives
William Alexander Percy Library (Greenville, Mississippi)

Books

Allen, Barbara. *Uneasy Alchemy: Citizens and Experts in Louisiana's Chemical Corridor Disputes*. Cambridge: MIT Press, 2003.

Allen, Michael. *Western Rivermen, 1763–1861: Ohio and Mississippi Boatmen and the Myth of the Alligator Horse*. Baton Rouge: Louisiana State University Press, 1990.

Andrews, Thomas. *Killing for Coal: America's Deadliest Labor War*. Cambridge: Harvard University Press, 2010.

Anfinson, John O. *The River We Have Wrought: A History of the Upper Mississippi River*. Minneapolis: University of Minnesota Press, 2003.

Argos, Collectif. *Climate Refugees*. Cambridge: MIT Press, 2010.

Arneson, E. *Waterfront Workers of New Orleans: Race, Class, and Politics*, 1863–1923. Urbana: University of Illinois Press, 1991.

Ayers, Edward L. *The Promise of the New South: Life After Reconstruction*. New York: Oxford University Press, 1992.

Baptist, Edward E. *The Half Has Never Told: Slavery and the Making of American Capitalism*. New York: Basic Books, 2014.

Barnett, James F., Jr. *Beyond Control: The Mississippi River's New Channel to the Gulf of Mexico*. Jackson: University Press of Mississippi, 2017.

Barry, John M. *Rising Tide: The Great Mississippi Flood of 1927 and How It Changed America.* New York: Simon & Schuster, 1997.

Bay, Mia. *The White Image in the Black Mind: African American Ideas about White People.* New York: Oxford University Press, 2000.

Beckert, Sven. *Empire of Cotton: A Global History.* New York: Vintage Books, 2014.

Berlin, Ira. *Many Thousands Gone.* Cambridge: Harvard University Press, 2000.

Blythe, Robert W. *Administrative History of Jean Lafitte National Historical Park and Preserve.* Washington: National Park Service, 2012.

Bolster, Jeffrey W. *Black-Jacks: African American Seamen in the Age of Sail.* Cambridge: Harvard University Press, 1998.

Bordewich, Fergus M. *Bound for Canaan: The Epic Story of the Underground Railroad, America's First Civil Rights Movement.* New York: Amistad, 2005.

Brinkley, Douglas. *The Great Deluge: Hurricane Katrina, New Orleans and the Mississippi Gulf Coast.* New York: Wm. Morrow and Company, 2006.

Buchanan, Thomas. *Black Life on the Mississippi: Slaves, Free Blacks, and the Western Steamboat World.* Chapel Hill: University of North Carolina Press, 2004.

Bullard, Robert D. *Dumping in Dixie: Race, Class, and Environmental Quality.* Oxford: Westview Press, 1990.

Bullard, Robert D. and Wright, Beverly. *Race, Place, and Environmental Justice after Hurricane Katrina: Struggles to Reclaim, Rebuild, and Revitalize New Orleans and the Gulf Coast.* Boulder: Westview Press, 2009.

Butler, Anne and Henderson, C. Murray. *Angola, Louisiana State Penitentiary a Half-Century of Rage and Reform.* Lafayette: University of Louisiana at Lafayette Press, 1990.

Camp, Stephanie. *Closer to Freedom: Enslaved Women and Everyday Resistance in the Plantation South.* Chapel Hill: University of North Carolina Press, 2004.

Carleton, Mark T. *Politics and Punishment: The History of the Louisiana State Penal System.* Baton Rouge: Louisiana State University Press, 1971.

Carney, Judith A. *Black Rice: The African Origins of Rice Cultivation in the Americas.* Cambridge: Harvard University Press, 2001.

Cioc, Marc. *The Rhine: An Eco-Biography,* 1815–2000. Seattle: University of Washington Press, 2002.

Coates, Peter. *A Story of Six Rivers: History, Culture, and Ecology.* London: Reaktion Books, 2013.

Cobb, James C. *The Most Southern Place on Earth: The Mississippi Delta and the Roots of Regional Identity.* New York: Oxford University Press, 1992.

Cole, Luke W. and Foster, Sheila R. *From the Ground Up: Environmental Racism and the Rise of the Environmental Justice Movement.* New York: New York University Press, 2001.

Colten, Craig E. *An Unnatural Metropolis: Wresting New Orleans from Nature.* Baton Rouge: Louisiana State University, 2005.

Comment, Bernard. *The Painted Panorama.* New York: Harry N. Abrams, Inc., Publishers, 2000.

Cone, James H. *Risks of Faith: The Emergence of a Black Theology of Liberation, 1968–1998.* Boston: Beacon Press, 1999.

Cowan, William Tynes. *The Slave and the Swamp: Disrupting the Plantation Narrative.* New York: Routledge, 2005.

Cowdrey, Albert E. *Land's End: A History of the New Orleans District, U.S. Army Corps of Engineers, and Its Lifelong Battle with the Lower Mississippi and Other Rivers Wending Their Way to the Sea.* Washington: United States Army Corps of Engineers, 1977.

Cusack, Tricia. *Riverscapes and National Identity.* Syracuse: Syracuse University Press, 2010.

Cutter, Susan, et al. *Hurricane Katrina and the Forgotten Coast of Mississippi.* New York: Cambridge University Press, 2014.

Daniel, Pete. *Deep'N As It Come: The 1927 Mississippi River Flood.* New York: Oxford University Press, 1977.

Daniel, Pete. *The Shadow of Slavery: Peonage in the South, 1901–1969.* Urbana: University of Illinois Press, 1972.

Davis, Francis. *The History of the Blues: The Roots, the Music, the People from Charley Patton to Robert Cray.* New York: Hyperion, 1995.

Davis, Thadious M. *Southscapes: Geographies of Race, Region & Literature.* Chapel Hill: University of North Carolina Press, 2011.

Dawdy, Shannon Lee. *Building the Devil's Empire: French Colonial New Orleans.* Chicago: University of Chicago Press, 2008.

Dawdy, Shannon Lee. *Patina: A Profane Archaeology.* Chicago: University of Chicago Press, 2016.

Diouf, Sylvia. *Slavery's Exiles: The Story of the American Maroons.* New York: New York University Press, 2014.

Dixon, Melvin. *Ride Out the Wilderness: Geography and Identity in Afro-American Literature.* Urbana: University of Illinois Press, 1987.

Donovan, Frank. *River Boats of America: From Flatboats to Floating Palaces, From the Cotton Blossom to the Staten Island Ferry.* New York: Thomas Y. Crowell Co., 1966.

DuBois, W.E.B. *Black Reconstruction in America, An Essay toward a History of the Part Which Black Folk Played in an Attempt to Reconstruct Democracy in America, 1860–1880.* 1935, New York: Free Press, rpt. 1997.

Dungy, Camille T., ed. *Black Nature: Four Centuries of African American Nature Poetry,* Athens: University of Georgia Press, 2009.

DuVal, Kathleen. *Independence Lost: Lives on the Edge of the American Revolution.* New York: Random House, 2016.

Dyson, Michael. *Come Hell or High Water: Hurricane Katrina and the Color of Disaster.* Davidson: Civitas Press, 2007.

Dyson, Michael Eric. *Come Hell or High Water: Hurricane Katrina and the Color of Disaster.* New York: Basic Books, 2005; rpt. 2007.

Ermus, Cindy, ed. *Environmental Disaster in the Gulf South: Two Centuries of Catastrophe, Risk, and Resilience.* Baton Rouge: Louisiana State University Press, 2018.

Evans, David. *Big Road Blues: Tradition and Creativity in the Folk Blues.* Berkeley: University of California, 1982.

Eyerman, Ron. *Cultural Trauma: Slavery and the Formation of African American Identity.* Cambridge: Cambridge University Press, 2001.

Finney, Carolyn. *Black Faces, White Spaces: Reimagining the Relationship of African Americans to the Great Outdoors.* Chapel Hill. University of North Carolina Press, 2014.

Follett, Richard. *The Sugar Masters: Planters and Slaves in Louisiana's Cane World* Baton Rouge. Louisiana State University Press, 2005.

Foner, Eric. *Reconstruction, Updated Edition: America's Unfinished Revolution, 1863–1877.* New York: Harper Perennial Modern Classics, 2014.

Franklin, John Hope and Schweninger, Loren. *Runaway Slaves: Rebels on the Plantation.* New York: Oxford University Press, 1999.

Frey, Sylvia. *Water from the Rock: Black Resistance in a Revolutionary Age.* Princeton: Princeton University Press, 1991.

Gara, Larry. *The Liberty Line: The Legend of the Underground Railroad.* Lexington: University Press of Kentucky, 1991.

Gershenhorn, Jerry. *Melville J. Herskovits and the Racial Politics of Knowledge*. Lincoln: University of Nebraska Press, 2004.

Glave, Diane D. *Rooted in the Earth: Reclaiming the African American Environmental Heritage*. Chicago: Lawrence Hill Books, 2010.

Gould, E.W. *Fifty Years on the Mississippi or Gould's History of River Navigation*. St. Louis: Nixon-Jones Printing Co., 1889.

Grant, Nancy. *TVA and Black Americans Planning for the Status Quo*. Philadelphia: Temple University Press, 1990.

Hall, Gwendolyn Mildo. *Africans in Colonial Louisiana: The Development of Afro-Creole Culture in the Eighteenth Century*. Baton Rouge: Louisiana State University Press, 1992.

Hammond, John Craig. "Slavery, Settlement, and the Empire: The Expansion and Growth of Slavery in the Interior of the North-American Continent, 1770–1820" *Journal of the Early Republic*, Vol. 32, No. 2 (Summer 2012):175–206.

Harris, Eddy L. *Mississippi Solo: A Memoir*. New York: Henry Holt and Company, 1988.

Horowitz, Andy. *Katrina: A History, 1915–2015*. Cambridge: Harvard University Press, 2020.

Issac, Rhys. *The Transformation of Virginia, 1740–1790*. Chapel Hill: University of North Carolina Press, 1999.

Jack, Bryan M. *The St. Louis African American Community and the Exodusters*. Columbia: University of Missouri Press, 2008.

Jackson, Joyce Marie. *Life in the Village: A Cultural Memory of the Fazendeville Community*. Washington: U.S. Department of the Interior, National Historical Park, 2003.

Johnson, Charles. *Growing Up in the Black Belt: Negro Youth in the Rural South*. Washington: American Council of Education, 1941.

Johnson, Rashuana. *Slavery's Metropolis: Unfree Labor in New Orleans during the Age of Revolution*. Cambridge: Cambridge University Press, 2016.

Johnson, Walter. *River of Dark Dreams: Slavery and Empire in the Cotton Kingdom*. Cambridge: The Belknap Press of Harvard University Press, 2013.

Joy, Mark. *American Expansionism, 1763–1860: A Manifest Destiny*. London: Taylor and Francis, 2014.

Kastor, Peter J. and Weil, Francois, eds. *Empires of the Imagination: Transatlantic Histories of the Louisiana Purchase*. Charlottesville: University of Virginia Press, 2009.

Kaye, Anthony E. *Joining Places: Slave Neighborhoods in the Old South*. Chapel Hill: University of North Carolina Press, 2007.

Kenney, William Howland. *Jazz on the River*. Chicago: University of Chicago Press, 2005.

Kmen, H. *Music in New Orleans: The Formative Years, 1791–1841*. Baton Rouge: Louisiana State University Press, 1996.

Krehbiel, Henry Edward. *Afro-American Folksongs: A Study in Racial and National Music*. New York: G. Schirmer, 1914.

LaFlouria, Tabitha. *Chained in Silence: Black Women and Convict Labor in the New South*. Chapel Hill: University of North Carolina Press, 2015.

Leman, Nicholas. *The Promised Land: The Great Black Migration and How It Changed America*. New York: Vintage Books, 1992.

Lerner, Steve. *Diamond: A Struggle for Environmental Justice in Louisiana's Chemical Corridor*. Cambridge: MIT Press, 2006.

Levine, Lawrence. *Black Culture and Black Consciousness*. Oxford: Oxford University Press, 1977; rpt. 2007.

Linebaugh, Peter and Rediker, Marcus. *The Many-Headed Hydra: Sailors, Slaves, Commoners, and the Hidden History of the Revolutionary Atlantic*. Boston: Beacon Press, 2000.

Lomax, Alan. *The Land Where the Blues Began*. New York: Pantheon Books, 1993.
Mancini, Matthew J. *"One Dies, Get Another": Convict Leasing in the American South, 1866–1928*. Columbia: University of South Carolina Press, 1996.
McDermott, John Francis. *George Caleb Bingham: River Portraitist*. Norman: University of Oklahoma Press, 1959.
McDonald, Roderick A. *The Economy and Material Culture of Slaves: Goods and Chattels on the Sugar Plantations of Jamaica and Louisiana*. Baton Rouge: Louisiana State University Press, 1993.
McKittrick, Katherine. *Demonic Grounds: Black Women and the Cartographies of Struggle*. Minneapolis: University of Minnesota Press, 2006.
McPherson, Tara. *Reconstructing Dixie: Race, Gender, and Nostalgia in the Imagined South*. Durham: Duke University Press, 2003.
Miller, David C. *Dark Eden: The Swamp in Nineteenth-Century American Culture*. Cambridge: Cambridge University Press, 1989.
Mizelle, Richard M. *Backwater Blues: The Mississippi Flood of 1927 in the African American Imagination*. Minneapolis: University of Minnesota Press, 2015.
Montrie, Chad. *Making a Living: Work and Environment in the United States*. Chapel Hill: University of North Carolina Press, 2008.
Morris, Christopher. *The Big Muddy: An Environmental History of the Mississippi and Its Peoples from Hernando De Soto to Hurricane Katrina*. New York: Oxford University Press, 2012.
Mostern, Ruth. *The Yellow River: A Natural and Unnatural History*. New Haven: Yale University Press, 2021.
New York Historical Society. *Hudson River School: Nature and the American Vision*. New York: Rizzoli Electa, 2009.
Nixon, Rob. *Slow Violence and the Environmentalism of the Poor*. Cambridge: Harvard University Press, 2011.
O'Daniel, Patrick. *When the Levee Breaks: Memphis and the Mississippi Valley Flood of 1927*. Charleston: The History Press, 2013.
Oliver, Paul. *Blues Fell This Morning: Meaning in the Blues*. 1960; rpt. Cambridge: Cambridge University Press, 1990.
Oshinsky, David M. *"Worse Than Slavery": Parchman Farm and the Ordeal of Jim Crow Justice*. New York: Free Press, 1996.
Ostendorf, Anne. *Sounds American: National Identity and the Music Culture of the Lower Mississippi River Valley, 1800–1860*. Athens: University of Georgia Press, 2011.
Outka, Paul. *Race and Nature: From Transcendentalism to the Harlem Renaissance*. New York: Palgrave Macmillan, 2013.
Painter, Nell Irwin. *Exodusters: Black Migration to Kansas*. New York: W.W. Norton and Company, 1992.
Parrish, Susan Scott. *The Flood Year 1927*. Princeton: Princeton University Press, 2017
Pitts, Walter F., Jr. *Old Ship of Zion: The Afro-Baptist Ritual in the African Diaspora*. New York: Oxford University Press, 1993
Price, Richard, ed. *Maroon Societies: Rebel Slave Communities in the Americas*, 3rd ed. Baltimore: Johns Hopkins University Press, 1996.
Pritchard, Sara. *Confluence: The Nature of Technology and the Remaking of the Rhone*. Cambridge: Harvard University Press, 2011.
Quick, Herbert and Quick, Edward. *Mississippi Steamboatin': A History of Steamboating on the Mississippi and Its Tributaries*. New York: Henry Holt and Company, 1926.
Raboteau, Albert J. *Slave Religion: The "Invisible Institution" in the Antebellum South*. New York: Oxford University Press, 2004.

Rakoff, Jed. *Why the Innocent Plead Guilty and the Guilty Go Free.* New York: Farrar, Straus & Girous, 2021.

Rasmussen, Daniel. *American Uprising: The Untold Story of America's Largest Slave Revolt.* New York: HarperCollins, 2011.

Rivlin, Gary. *Katrina: After the Flood.* New York: Simon & Schuster, 2015.

Roberts, John W. *From Trickster to Badman: The Black Folk Hero in Slavery and Freedom.* Philadelphia: University of Pennsylvania Press, 1989.

Rogers, Thomas D. *The Deepest Wounds: A Labor and Environmental History of Sugar in Northeast Brazil.* Chapel Hill: University of North Carolina Press, 2010.

Rothman, Adam. *Slave Country: American Expansion and the Origins of the Deep South.* Cambridge: Harvard University Press, 2005.

Ruffin, Kimberly N. *Black on Earth: African American Ecoliterary Traditions.* Athens: University of Georgia Press, 2010.

Said, Edward. *Culture and Imperialism.* London: Chatto & Windus, 1993.

Saikku, Mikko. *This Delta, This Land: An Environmental History of the Yazoo-Mississippi Floodplain.* Athens: University of Georgia Press, 2005.

Sandlin, Lee. *Wicked River: The Mississippi When It Last Ran Wild.* New York: Vintage Books, 2010.

Savoy, Lauret. *Trace: Memory, History, Race, and the American Landscape.* Berkeley: Counterpoint, 2015.

Sayers, Daniel. *A Desolate Place for a Defiant People: The Archaeology of Maroons, Indigenous and Enslaved Laborers in the Great Dismal Swamp.* Gainesville: University Press of Florida, 2014.

Schneider, Paul. *Old Man River: The Mississippi in North American History.* New York: Henry Holt and Company, 2013.

Sciolino, Elaine. *The Seine: The River That Made Paris.* New York: W.W. Norton and Company, 2019.

Seelye, John. *Prophetic Waters: The River in Early American Life and Literature.* New York: Oxford University Press, 1977.

Shallat, Todd. *Structures in the Stream: Water, Science, and the Rise of the U.S. Army Corps of Engineers.* Austin: University of Texas Press, 1994.

Smith, John David, ed. *Reconstruction: Interpreting American History.* Kent: Kent University Press, 2016.

Smith, Kimberly. *African American Environmental Thought.* Lawrence: University Press of Kansas, 2007.

Smith, Thomas Ruys. *Deep Water: The Mississippi River in the Age of Mark Twain.* Baton Rouge: Louisiana State University Press, 2019.

Smith, Thomas Ruys. *River of Dreams: Imagining the Mississippi before Mark Twain.* Baton Rouge: Louisiana State University Press, 2007.

Springer, Robert, ed. *Nobody Knows Where the Blues Come From: Lyrics and History.* Jackson: University Press of Mississippi, 2006.

Stewart, Mart. *What Nature Suffers to Groe: Life, Labor, and Landscape on the Georgia Coast, 1680–1920.* Athens: University of Georgia Press, 2002.

Stoll, Mark R. *Inherit the Mountain: Religion and the Rise of American Environmentalism.* New York: Oxford University Press, 2015.

Sublette, Ned. *The World that Made New Orleans: From Spanish Silver to Congo Square.* Chicago: Lawrence Hill Books, 2008.

Taylor, Dorceta E. *Toxic Communities: Environmental Racism, Industrial Pollution, and Residential Mobility.* New York: New York University Press, 2014.

Trotter, Joe William, Jr. *River Jordan: African American Urban Life in the Ohio Valley.* Lexington: University Press of Kentucky, 1998.

Troutt, David Dante, ed. *After the Storm: Black Intellectuals Explore the Meaning of Hurricane Katrina.* New York: The New Press, 2007.

Usner, Daniel. *Indians, Settlers, and Slaves in a Frontier Exchange Economy: The Lower Mississippi Valley before 1783.* Chapel Hill: University of North Carolina Press, 1992.

Videl, Cecile, ed. *Louisiana: Crossroads of the Atlantic World.* Philadelphia: University of Pennsylvania Press, 2014.

Vileisis, Ann. *Discovering the Unknown Landscape: A History of America's Wetlands.* Washington: Island Press, 1997.

Vlach, John M. *Back of the Big House: The Architecture of Plantation Slavery.* Chapel Hill: University of North Carolina Press, 1993.

Wailoo, K. and O'Neil, Kevin M., et al., eds. *Katrina's Imprint: Race and Vulnerability in America.* Rutgers Studies on Race and Ethnicity. New Brunswick: Rutgers University Press, 2010.

Wennersten, John and Robbins, Denise. *Rising Tides: Climate Refugees in the Twenty-First Century.* Bloomington: Indiana University Press, 2017.

Wharton, Vernon Lane. *The Negro in Mississippi, 1865–1890.* New York: Harper and Row, 1947.

White, Deborah Gray. *Ain't I a Woman?: Female Slaves in the Plantation South.* New York: W.W. Norton & Co., 1985.

White, Shane and White, Graham. *The Sounds of Slavery: Discovering African American History through Songs, Sermons, and Speech.* Boston: Beacon Press, 2005.

Willis, John C. *Forgotten Time: The Yazoo-Mississippi Delta after the Civil War.* Charlottesville: University Press of Virginia, 2000.

Woodward, C. Vann. Origins of the New South: 1877–1913. Baton Rouge: Louisiana State University Press, 1971.

Zeisler-Vralsted, Dorothy. Rivers, Memory, and Nation-Building: A History of the Volga and Mississippi Rivers. New York: Berghahn Books, 2014.

Articles, Chapters, Dissertations, and Theses

Abrams, Eve. "The Defeat of Fazendeville" *64 Parishes,* Louisana Endowment for the Humanities, n.d.

Allen, Barbara. "Cradle of a Revolution? The Industrial Transformation of Louisiana's Lower Mississippi River" *Technology and Culture,* Vol. 47, No. 1 (January 2006):112–119.

Antonellis, April. "New Orleans and Fazendeville (de)Segregated: Challenging a Narrative of School Integration" MA Thesis, Louisiana State University, 2012.

Aptheker, Herbert. "Maroons within the Present Limits of the United States" *The Journal of Negro History,* Vol. 24, No. 2 (April 1939):167–184.

Barca, Stephania. "Laboring the Earth: Transnational Reflections on the Environmental History of Work" *Environmental History,* Vol. 19, No. 1 (2014):3–27.

Baugh, Joyce. "African-Americans and the Environment: A Review Essay" *Policy Studies Journal,* Vol. 19, No. 2 (September 2005):182–191.

Berry, Gregory E. "Organizing against Multinational Corporate Power in Cancer Alley: The Activist Community as Primary Stakeholder" *Organization and Environment,* Vol. 16, No. 1 (March 2003):3–33.

Blake, Lee Ella. "The Great Exodus of 1879 and 1880 to Kansas" MA Thesis, Kansas State College, 1942.

Blassingame, John W. "Using the Testimony of Ex-Slaves: Approaches and Problems" *Journal of Southern History*, Vol. 41, No. 4 (November 1975):473–492.

Blight, David. "The Slave Narratives: A Genre and a Source" *History Now: The Journal of the Gilder Lehrman Institute* accessed at https://www.gilderlehrman.org/history-by-era/literature-and-language-arts/essays/slave-narrative.

Blum, Elizabeth D. "Power, Danger, and Control: Slave Women's Perceptions of Wilderness in the Nineteenth Century" *Women's Studies*, Vol. 31 (2002):247–265.

Buchanan, Thomas. "Levees of Hope: African American Steamboat Workers, Cities, and Slave Escapes on the Antebellum Mississippi" *Journal of Urban History*, Vol. 30, No. 3 (March 2004):360–377.

Buchanan, Thomas. "Rascals on the Antebellum Mississippi: African American Steamboat Workers and the St. Louis Hanging of 1841" *Journal of Social History*, Vol. 34, No. 4 (July 2001):797–816.

Bullard, Robert D. "The Threat of Environmental Racism" *Natural Resources & Environment*, Vol. 7, No. 3 (Winter 1993):23–26.

Byrnes, Malcolm. "Climate Justice, Hurricane Katrina, and African American Environmentalism" *Journal of African American Studies*, Vol. 18, No. 3 (September 2014):305–314.

Cade, John B. "Out of the Mouths of Ex-Slaves" *Journal of Negro History*, Vol. 20, No. 3 (July 1935):294–337.

Cardon, Nathan. ""Less than Mayhem": Louisiana's Convict Lease, 1865–1901" *Louisiana History*, Vol. 58, No. 4 (Fall 2017):417–441.

Carroll, Ginny. "When Pollution Hits Home" *National Wildlife Federation* (1 August 1991).

Chakborty, J., et al. "Environmental Justice Research: Contemporary Issues and Emerging Topics" *International Journal of Environmental Research and Public Health* (1 November 2016) accessed at https://doi.org/10.3390/ijerph13111072/.

Colten, Craig E. "An Incomplete Solution: Oil and Water in Louisiana" *Journal of American History*, Vol. 99, No. 1 (June 2012):91–99.

Colten, Craig E. "The Rusting of the Chemical Corridor" *Technology and Culture*, Vol. 47, No.1 (January 2006):95–101.

Colten, Craig E. "Vulnerability and Place: Flat Land and Uneven Rise in New Orleans" *American Anthropologist*, Vol. 108, No. 4 (December 2006):731–734.

Crowley, J. "Shack Bullies and Levee Contractors: Bluesmen as Ethnographers" *Journal of Folklore Research*, Vol. 28, No. 2 (1991):135–162.

Dawdy, Shannon Lee. "The Burden of Louis Congo and he Evolution of Savagery in Louisiana" *Discipline and the Other Body: Correction, Corporeality, Colonialism*, ed. Steven Pierce and Anupama Rao (Durham: Duke University Press, 2006):61–90.

DeLuca, Kevin. "In the Shadow of Whiteness: The Consequences of Constructions of Nature in Environmental Politics" *Whiteness: The Communication of Social Identity*, eds. Thomas K. Nakayama and Judith N. Martin (Thousand Oaks, CA: Sage, 1999):217–245.

DENÉVE, Stéphanie. "Representing Environmental Emergency as Social Emergency: The Great Mississippi Flood of 1927 in Blues Songs from Louisiana and Mississippi" *Catalogue Des 559 Revues* (2 August 2021).

Din, Gilbert. ""Cimarrones" and the San Malo Band in Spanish Louisiana" *Louisiana History: The Journal of the Louisiana Historical Society*, Vol. 21, No. 3 (Summer 1980):247–248.

du Villiers, Baron Mark. "A History of the Foundation of New Orleans (1717–1722)" *Louisiana Historical Quarterly*, Vol. 3, (January-October, 1920):157.

Dyl, Joanna. "Transience, Labor and Nature: Itinerant Workers in the American West" *International Labor and Working Class History*, Vol. 85 (April 2014):97–117.

Eltis, D., Morgan, P., and Richardson, D. "Agency and Diaspora in Atlantic History: Reassessing the African Contribution to Rice Cultivation in the Americas" *American Historical Review*, Vol. 112, No. 5 (December 2007):1329–1358.

Fessenden, Marissa. "How a Nearly Successful Slave Revolt Was Intentionally Lost to History" *SmithsonianMag.com*, January 8, 2016 accessed at https://www.smithonian-mag.com/smart-news/its-anniversary-1811-louisian-slave-revolt-180957760/.

Fraser, Steve and Freeman, Joshua B. "In the Rearview Mirror: Barbarism and Progress, the Story of Convict Labor" *New Labor Forum*, Vol. 21, No. 3 (Fall 2017):90–92.

Fussell, Elizabeth. "Constructing New Orleans, Constructing Race: A Population History of New Orleans" *Journal of American History*, Vol. 94 (December 2007):846–855.

Fussell, Elizabeth. "The Long-Term Recovery of New Orleans' Population after Hurricane Katrina" *American Behavioral Scientist*, Vol. 59, No. 10 (September 2015):1231–1245.

Gentile, Phillip "Viewing the Iconic Mississippi: Strategies of Reenactment in River Panoramas and Bill Morrison's *The Great Flood* (2013)" *Southern Quarterly*, Vol. 53, No. 3 (Spring 2015):121–136.

George, Courtney. "From Bounce to the Mainstream: Hip Hop Representations of Post-Katrina New Orleans in Music, Film, and Television" *European Journal of American Culture*, Vol. 35, No. 1 (2016):17–32.

Giancarlo, Alexandra. "The Lower Ninth Ward: Resistance, Recovery, and Renewal" MA Thesis, Louisiana State University, 2011.

Giltner, Scott. "Slave Hunting and Fishing in the Antebellum South" *To Love the Wind and the Rain: African Americans and Environmental History*, eds. Dianne Glave and Mark Stoll (Pittsburgh: University of Pittsburgh Press, 2005):21–37.

Ginsberg, Rebecca. "Freedom and the Slave Landscape" *Landscape Journal*, Vol. 26 (2007):1–7.

Gooch, Cynthia. "I've Known Rivers: Representations of the Mississippi River in African American Literature and Culture" Ph.D. Diss., University of Kentucky, 2019.

Haas, Edward F. ""Don't Believe Any False Rumors . . ." Mayor Victor H. Schiro, Hurricane Betsy and Urban Myths" *Louisiana History*, Vol. 45, No. 4 (Autumn 2004):463–468.

Hall, Gwendolyn and Rowell, Charles Henry. "American Tragedy: New Orleans under Water" *Callaloo*, Vol. 29, No. 4 (2006):1049–1055.

Hannah-Jones, Nikole. "The 1619 Project" *New York Times Magazine* (14 August 2019)

Hermann, Christina Pruett. "Specters of Freedom: Forced Labor, Social Struggle, and the Louisiana Penitentiary System, 1835–1935" Ph.D. Diss. Michigan State University, 2015.

Hines, Alicia. "Geographies of Freedom: Black Women's Mobility and the Making of the Western River World, 1814–1865" Ph.D. Diss. Duke University, 2018.

Horowitz, Andy. "Hurricane Betsy and the Politics of Disaster in New Orleans" *Journal of Southern History*, Vol. 80, No. 4 (November 2014):893–934.

Houston, Donna. "Crisis is Where We Live: Environmental Justice for the Anthropocene" *Globalizations*, Vol. 10, No. 3 (July 2013):439–450.

Jackson, Joyce. "Declaration of Taking Twice: The Fazendeville Community of the Lower Ninth Ward Empowering Place" *American Anthropologist*, Vol. 108, No. 4 (December 2006):766–780.

Jordan, Rosan Augusta and De Caro, Frank. ""In this Folk-Lore Land": Race, Class, Identity, and Folklore Studies in Louisiana" *The Journal of American Folklore*, Vol. 109, No. 431 (Winter 1996):31–59.

Kelman, Ari. "In the Shadow of Disaster" *The Nation* (15 December 2005) accessed at https://www.thenation.com/article/archive/shadow-disaster/.

Klinkenberg, Dean. "The 70 Million-Year-Old History of the Mississippi River" *Smithsonian Magazine* (September 2020) accessed at https://www.smithsonianmag.com/science-nature/geological-history-mississippi-river-180975509/.

LaFlouria, Tabitha. ""Under the Sting of the Lash": Gendered Violence, Terror, and Resistance in the South's Convict Camps" *Journal of American History,* Vol. 100, No. 3 (Summer 2015):366–384.

Landphair, Juliette. ""The Forgotten People of New Orleans": Community, Vulnerability, and the Lower Ninth Ward" *Journal of American History*, 94 (December 2007). Accessed at https://doi.org/10.2307/25095146.

Lavin, Chavid and Russell, Chris. "The Buoyancy of Failure: Battling Nature in New Orleans" *Space and Culture*, Vol. 9, No. 1 (February 2006):48–51.

LeMenager, Stephanie. "Floating Capital: The Trouble with Whiteness on Twain's Mississippi" *ELH*, Vol. 71, No. 2 (Summer 2004):405–431.

Lovell, John, Jr. "The Social Implications of the Negro Spiritual" *Journal of Negro Education*, Vol. 8, No. 4 (October 1939):634–643.

McCoyer, Michael. ""Rough Mens" in the Toughest Places I Have Ever Seen: The Construction and Ramifications of Black Masculine Identity in the Mississippi Delta's Levee Camps, 1900–1935" *International Labor and Working-Class History*, No. 69, Issue 1 (Spring 2006):57–80.

McGowan, James. "Creation of a Slave Society: Louisiana Plantations in the Eighteenth Century" Ph.D. Diss., University of Rochester, 1976.

McGurty, Eileen Maura. "From NIMBY to Civil Rights: The Origins of the Environmental Justice Movement" *Environmental History*, Vol. 2, No. 3 (July 1997):301–323.

McKee, Seth C. "Politics in Black and White: The Mississippi Delta" *Defining the Delta: Multidisciplinary Perspectives on the Lower Mississippi River Delta*, ed. Janelle Collins (Fayetteville: University of Arkansas Press, 2015:133–144.

McKelvey, Blake. "A Half Century of Southern Penal Exploitation" *Social Forces*, Vol. 13, No. 1 (October 1934–May 1935):112–113.

McKibben, Davidson Burns. "Negro Slave Insurrections in Mississippi, 1800–1865" *The Journal of Negro History*, Vol. 34, No. 1 (January, 1949):73–90.

McMurchy, Myles. ""The Red Cross is Not All Right!": Herbert Hoover's Concentration Cover-Up in the 1927 Mississippi Flood" *Yale Historical Review* 5 (Fall 2015):87–114.

Merchant, Carolyn. "Shades of Darkness: Race and Environmental History" *Environmental History*, No. 8 (July 2003):380–394.

Mizelle, Richard L. "Black Levee Camp Workers, the NAACP, and the Mississippi Flood Control Project, 1927–1933" *The Journal of African American History,* Vol. 98, No. 4 (Fall 2013):511–530.

Moore, John Hebron. "Simon Gray, Riverman: A Slave Who Was Almost Free" *Mississippi Valley Historical Review*, Vol. 49, No. 3 (December, 1962):472–484.

Moore, John Hebron. "Two Cotton Kingdoms" *Agricultural History,* Vol. 60, No. 5 (Autumn 1986):1–16.

Nuwer, Rachel. "Hurricane Katrina and Rita Caused at Least 117 Uncounted Deaths" *Smithsonian Magazine* (9 May 2014) accessed at https://www.smithsonianmag.com/smart-news/hurricanes-katrina-and-ritas-death- tolls-increase-when-stillbirths-are-taken-account-180951388/.

Paul,Richard."IntheFieldofFolkMusic,AlanLomaxisaGiant—IfaFlawedandControversial One" *The World* (February 2015) accessedat https:///www.pri.org/stories/2015-02-10/field-folk-muic-alan-lomax-giant-if-flawed-and-controversial-one

Peck, Gunther. "The Nature of Fault Lines and Common Ground in Environmental and Labor History" *Environmental History*, Vol. 11, No. 2 (2006):212–238.

Peek, Lori, et al. "Disaster Hits Home: A Model of Displaced Family Adjustment after Hurricane Katrina" *Journal of Family Issues*, Vol. 32, No. 10 (2011):1371–1396.

Pellow, David. "Toward a Critical Environmental Justice Studies: Black Lives Matter as an Environmental Justice Challenge" *Du Bois Review—Social Science Research on Race*, Vol. 13, No. 2 (October 2016):221–236.

Pena, Allison. "Wade in the Water: Personal Reflections on a Storm, a People, and a National Park" *American Anthropologist*, Vol. 108, No. 4 (December 2006):781–798.

Rivera, Jason David and Miller, DeMond Shondell. "Continually Neglected: Situating Natural Disasters in the African American Experience" *Journal of Black Studies*, Vol. 34, No. 4 (March 2007):502–522.

Sensbach, Jon F. "The Singing of the Mississippi: The River and Religions of the Black Atlantic" *Religion in North America: Gods of the Mississippi*, ed. Michael Pasquier (Bloomington: Indiana University Press, 2013):17–36.

Sinclair, Bryan T. "Merging Streams: The Importance of the River in the Slaves' Religious World" *Journal of Religious Thought*, Vol. 53, No. 2 (1997):1–19.

Smith, John David. "The Unveiling of Slave Folk Culture, 1865–1920" *Journal of Folklore Research*, Vol. 21, No. 1 (April 1984):47–56.

Smith, Thomas Ruys. ""The Mississippi Was a Virgin Field": Reconstructing the River Before Mark Twain, 1865–1875" *Mark Twain Journal*, Vol. 53, No. 2 (Fall 2015):24–66.

Spencer, Robyn. "Contested Terrain: The Mississippi Flood of 1927 and the Struggle to Control Black Labor" *The Journal of Negro History,* Vol. 79, No. 2 (Spring 1994):170–181.

Stewart, Mart. "If John Muir Had Been an Agrarian: American Environmental History West and South" *Environment and History*, Vol. 11, No. 2 (May 2005):139–162.

Surrey, N.M. "The Commerce of Louisiana during the French Regime, 1699–1768" Ph.D. Diss., Columbia University, 1916.

Thompson, Shirley. "The Long View from the Levee" *Transition,* No. 108 (2012):1–17

Thompson, Thomas Marshall. "National Newspapers and Legislative Reactions to Louisiana's Deslondes Slave Revolt of 1811" *The Journal of the Louisiana Historical Association*, Vol. 33, No. 1 (Winter 1992):5–29.

Thorpe, Thomas Bangs. "Remembrances of the Mississippi" *Harpers New Monthly Magazine,* Vol. 12 (December 1855–May 1856):25–27.

Thwaites, Reuben Gold, ed. "Letter from Father du Poisson" *Travels and Explorations of the Jesuit Missionaries in New France, 1610–1791, Vol. LXVII* (Cleveland: The Burrows Brothers Company, 1900):289

Usner, Daniel H., Jr. "From African Captivity to American Slavery" *Louisiana History*, Vol. 20, No. 1 (Winter 1979):165–192.

Welk, Terry. "The Archaeology of Maroon Societies in the Americas: Resistance, Cultural Continuity, and Transformation in the African Diaspora" *Historical Archaeology*, Vol. 31, No. 2 (1997):81–92.

White, Richard. "Are You an Environmentalist or Do You Work for a Living?" *Uncommon Ground: Rethinking the Human Place in Nature*, ed. William Cronon (New York: W.W. Norton & Company, 1996):171–185.

Whitmore, Janet L. "A Panorama of Unequaled Yet Ever-Varying Beauty" *Currents of Change: Art and Life along the Mississippi River, 1850–1861*, eds. Jason Busch and Christopher Monkhouse (Minneapolis: Minneapolis Institute of the Arts, 2004):12–62.

Woodruff, Nancy Elizabeth. *American Congo: The African American Freedom Struggle in the Delta.* Cambridge: Harvard University Press, 2003.

Wright, Richard. "The Man Who Saw the Flood" *Eight Men* (1940; rpt. New York: World Publishing, 1961):102–155.

Newspapers, Magazines, Newscasts

Atlantic Monthly
Chicago Defender
Commercial Appeal (Memphis)
The Crisis
Daily Democrat Times (Greenville)
The Guardian
Harper's
Los Angeles Times
Louisiana Courier
Louisiana Gazette
Metropolitics
Missouri Republican
The Nation
National Public Radio
New Orleans City Business
New Orleans Daily Picayune
The New Yorker
New York Times
New York Times Magazine
Republic (St. Louis)
St. Louis Globe Democrat
The Times (London)
UN News
The Washington Post

Government Documents

Mississippi Provincial Archives, 1701–1729
Senate Document N. 2, 39th Congress, 2nd Sess. (1865)
Senate Report No. 693, 46th Congress, 2nd Sess. (1880)
Spanish Judicial Records of Louisiana
U.S. Army Corps of Engineers. *Mississippi River Commission History* (June 2018)
U.S. General Services Administration. *Hard Labor: History and Archaeology at the Old Louisiana State Penitentiary* (1991)

Primary Sources

Alford, Terry. *Prince among Slaves: The True Story of an African Prince Sold into Slavery in the American South*. New York: Oxford University Press, 1977, rpt. 2007.

American National Red Cross. *The Final Report of the Colored Advisory Commission Appointed to Cooperate with the American National Red Cross and the President's Committee on Relief Work in Mississippi Valley Flood Disaster of 1927*. Washington: American National Red Cross, 1929.

American National Red Cross. *The Mississippi Valley Flood Disaster of 1927: Official Report of the Relief Operations.* Washington: American National Red Cross, 1928.

Anderson, William J. *Life and Narrative of William J. Anderson, Twenty-Four Years a Slave.* Chicago: Daily Tribune Book and Job Printing Office, 1857; rpt. 2016.

Bibb, Henry. *Narrative of the Life and Adventures of Henry Bibb, an American Slave, Written by Himself.* New York, 1849; rpt. Echo Library, 2005.

Blake, Henry. Interview. Federal Writers' Project: Slave Narrative Project, Vol. 2, Part 1, Arkansas (nd):Image 183 accessed at https://www.loc.gov/resource/mesn.021/?sp=180&st=text.

Blassingame, John W. *Slave Testimony: Two Centuries of Letters, Speeches, Interviews and Autobiographies.* Baton Rouge: Louisiana State University Press, 1977.

Boyer, Lucy. Interview. *Lower Mississippi Valley Collections. Bremer, Frances. The Homes of the New World—Impressions of America.* New York: Harper Brothers, 1853.

Brown, Lewis. Interview. Federal Writers project, Arkansas Slave Narratives (December 1938; rpt. 2007):Image 297.

Brown, William W. *The Narrative of William W. Brown: A Fugitive Slave.* 1847; rpt. Read-aClassic, 2010.

Bryant, William Cullen, ed. *Picturesque America.* New York: D. Appleton & Co., 1872, 1874.

Burns, George Taylor. "Interview" *Federal Writers Project, Slave Narrative Project*, Vol. 5: 37–39 accessed at https://www.loc.gov/resource/mesn.050/?sp=43.

Byrd, William. Description of the Dismal Swamp and a Proposal to Drain the Swamp (1728). Accessed at https://www.loc.gov/resource/lhbcb.22884/?sp=16.

Cable, George. "Creole Slave Songs" *Century Magazine*, 1886, p. 14, UNC digital https://archive.org/details/creoleslavesongs00cable/page/814.

Cable, George. *Silent South: Together with the Freedmen's Case in Equity and the Convict Lease System.* (1885) London: Forgotten Books, rpt. 2017.

Clayton, Lawrence Al, et al, ed. *The De Soto Chronicles, the Expedition of Hernando de Soto to North American in 1539–1543*, vol. 1. Tuscaloosa: University of Alabama Press, 1993.

Cohn, David. *Where I Was Born and Raised.* Notre Dame: University of Notre Dame Press, 1935.

Davis, William E. and Kiernan, F. Mark, eds. "Biographical Sketch of the Life and Ancestry of George Forman of Stratford-Ontario-Canada" in *The Mighty Mississippi: Two 19th Century Accounts. Louisiana History: The Journal of the Louisiana Historical Association*, Vol. 18, No. 3 (1977):338–348.

Davis, Winchester. Interview. Oral History Collection, Percy Library (Greenville). Douglass, Frederick. Narrative of the Life of Frederick Douglass, An American Slave. (1845) New York: Penguin Books, rpt. 1983.

Drew, Benjamin. *A North-Side View of Slavery: The Refugee: or, The Narratives of Fugitive Slaves in Canada, Related by Themselves, with an Account of the History and Condition of the Colored Population of Upper Canada*, Boston: J. P. Jewett & Co., 1856; rpt. General Books LLC, 2012.

Dumont, M. Some Abstracts from the Historical Memoirs of Louisiana, Including the Most Interesting Events from 1687 to the Present (1753) accessed at https://www.gutenberg.org/files/9153/9153-h/9153-h.htm#book-II-chapter-XII.

Flint, Timothy. *A Condensed Geography and History of the Western States, or the Mississippi Valley*, Vol. 1. Cincinnati: W.M. and O. Farnsworth, Jun. Printers, 1828.

Flint, Timothy. *The History and Geography of the Mississippi Valley*, Vol. 1, 3rd ed. Boston: Carter, Hendee and Co., 1833.

Fortman, George. Interview. Federal Writers' Project: Slave Narrative Project, Vol. 5, Indiana (nd):Image 88 accessed at https://www.loc.gov/resource/mesn.050/?sp=88.

Gardner, Willie D. Interview. Oral History Collection. Percy Library (Greenville).

Greene, Daisy Miller. Interview. Oral History Collection. Percy Library (Greenville).

Hamilton, Mary. *Trials of the Earth: The Autobiography of Mary Hamilton*. Oxford: University of Mississippi Press, 1992.

Hearn, Lafcadio. *Children of the Levee*. 1876, Lexington: University of Kentucky Press, rpt. 1957.

Henderson, Celia. Interview. Federal Writers' Project: Slave Narrative Project, Vol.12, Ohio, (nd):42 accessed at https://www.loc.gov/resource/mesn.120/?sp=46.

Henson, Josiah. *Father Henson's Story of His Own Life*. n.p., 1858; reprint 2016.

Higginson, Thomas Wentworth. "Negro Spirituals" *Atlantic Monthly*, Vol. 116 (June 1867):685–694.

Hodges, John O. *Delta Fragments: The Recollections of a Sharecroppers Son*. Knoxville: University of Tennessee Press, 2013.

Hughes, Langston. *The Collected Works of Langston Hughes*. Ed. Christopher C. DeSantis. Columbia: University of Missouri Press, 2001.

Jackson, Andrew. *Narrative and Writings of Andrew Jackson, of Kentucky*. Syracuse: Daily and Weekly Star Office, 1847; rpt. 2015.

Jackson, Bruce, ed. *The Negro and His Folklore in Nineteenth-Century Periodicals*. Austin: University of Texas Press, 1967.

Johnson, Clifton H. *God Struck Me Dead: Voices of Ex-Slaves*. Eugene: Wipf & Stock, 1969.

Joshua, Ida. Interview. Lower Mississippi Valley Collections.

Lawson, Ben. Interview. Federal Writers' Project: Slave Narrative Project, Vol. 13, Oklahoma, November 5, 1937: Image 182 accessed at https://www.loc.gov/resource/mesn.130/?sp=181.

Longfellow, Henry Wadsworth. *Henry Wadsworth Longfellow: Poems and Other Writings, Library of America Series*. New York: Library of America, 2000.

Mayers, Joe. Interview. Federal Writers' Project: Slave Narrative Project, Vol. 2, Part 5, Arkansas (nd): Image 74 accessed at https://www.loc.gov/resource/mesn.025/?sp=73.

McBride, Jesse Lee. Interview. Oral History Collection, Percy Library (Greenville). McClellan, George Marion. "A September Night" (1916) accessed at https://poets.org/poem/september-night.

McKaye, James. *The Mastership and Its Fruits: The Emancipated Slave, Face to Face with His Old Master (Valley of the Lower Mississippi*. New York: Loyal Publication Society, 1864 accessed at https://www.loc.gov/resource/lcrmrp.t2313/?sp=12&4=-0.398,0.177,1.813, 0.945,0.

Merrick, George. *Old Times on the Upper Mississippi: The Recollections of a Steamboat Pilot from 1854 to 1863, 1909*. Minneapolis: Minnesota Historical Society, 1987.

Murphy, Sally. Interview. Federal Writers' Project: Slave Narrative Project, Vol. 1, Alabama (1937):291 accessed at https://www.loc.gov/resource/mesn.010/?sp=300&st=-text; Northup, Solomon. Twelve Years a Slave. 1853, Auburn: Derby and Miller, rpt. 2014.

Olmsted, Frederick Law. *The Slave States*. New York: Capricorn Books, 1856, rpt. 1959.

Parker, John P. *His Promised Land: The Autobiography of John P. Parker, Former Slave and Conductor on the Underground Railroad*. New York: W.W. Norton, 1998.

Penner, D'Ann R. Overcoming Katrina: African American Voices from the Crescent and beyond. New York: Palgrave MacMillan, 2009.

Powdermaker, Hortense. *After Freedom: A Cultural Study in the Deep South*. New York: Viking Penguin, 1939; rpt. 1993.

Randolph, Peter. *Sketches of Slave Life from Slave Cabin to the Pulpit*. 1855. Ed. Katherine Clay Bassard. Morgantown: West Virginia University Press, rpt. 2016.

Rivera, Jason David and Miller, DeMond Shondell. "Continually Neglected: Situating Natural Disasters in the African American Experience" *Journal of Black Studies*, Vol. 34, No. 4 (March 2007):502–522.

Robin, C.C. *Voyage to Louisiana*. (1806) Gretna: A Firebird Press Book, 2000.

Ruffin, Thomas. *The Papers of Thomas Ruffin*. Raleigh: Edwards & Broughton Printing Co., 1918.

Schweninger, Loren, ed. *From Tennessee Slave to St. Louis Entrepreneur: The Autobiography of James Thomas*. Columbia: University of Missouri Press, 1984.

Scott, Janie. Interview. Federal Writers' Project: Slave Narrative Project, Vol. 12, Alabama (July 14, 1937):338 accessed at https://www.loc.gov/resource/mesn.010/?sp.

Sealsfield, Charles. The Americans as They Are: Described in a Tour through the Valley of the Mississippi, London: Bradbury and Dent, 1828. Accessed at http://www.gutenberg.org/files/44268/44268-h/44268-h.htm.

Simms, David McD. "The Negro Spiritual: Origins and Themes" *The Journal of Negro Education*, Vol. 35, No. 1 (Winter 1966):35–41

Slave Songs of the United States. New York: A. Simpson & Co., 1867; London: Forgotten Books, rpt. 2018.

Smith, Anna. Interview. Federal Writers' Project: Slave Narrative Project, Vol. 12, Ohio (11 June 1937):85 accessed at https://www.loc.gov/resource/mesn.120/?sp=89.

Smith, Barry. Interview. Federal Writers project. Mississippi Slave Narratives. (1936–38; rpt. 2007).

Southern Alluvian Land Association. The Call of the Alluvial Empire: Containing Authentic Information about the Alluvial Region of the Lower Mississippi Valley, Particularly the States of Arkansas, Tennessee, Mississippi and Louisiana. Southern Alluvian Land Association, 1919.

Stier, Issac. Interview. Federal Writers' Project: Slave Narrative Project, Vol. 9, Mississippi (1936–38):143 accessed at https://www.loc.gov/resource/mesn.090/?sp=15.

Stone, William. Interview. Federal Writers' Project: Slave Narrative Project, Vol. 16, Part 4, Texas (nd):Image 72 accessed at https://www.loc.gov/resource/mesn.164/?sp=70&st=text.

Tanner, Lynette Ater. *Chained to the Land: Voices from Cotton & Cane Plantations*. Winston-Salem: John F. Blair, Publisher, 2014.

Taylor, Jacob N. Sketch Book of Saint Louis: Containing a Series of Sketches of the Early Settlement, Public Buildings, Hotels, Railroads, etc. Missouri State Historical Society. Thomas, Omelia. Interview. Federal Writers' Project: Slave Narrative Project, Vol. 2, Part 6, Arkansas, (nd): Image 305 accessed at https://www.loc.gov/resource/mesn.026/?sp=305.

Trowbridge, J.T. *The South*. Hartford: L. Stebbins, 1866.

Watson, Henry. *Narrative of Henry Watson: A Fugitive Slave*. Boston: Bela Marsh, 1850.

Webb, Ishe. Interview. Federal Writers' Project: Slave Narrative Project, Vol. 2, Part 7, Arkansas, (nd):Image 80 accessed at https://www.loc.gov/resource/mesn.027/?sp=85.

Wheeler, Mary. *Steamboatin' Days. Folk Songs of the River Packet Era*. Baton Rouge: Louisiana State University Press, 1944.

White, Walter. "I Investigate Lynchings" *American Mercury* (January 1929).

White, Walter. "The Negro and the Flood" *The Nation*, Vol. 124, No. 3233 (22 June 1927):688–689.

Wilson, T. *1858 Interview in Slave Testimony: Two Centuries of Letters, Speeches, Interviews and Autobiographies*, ed. J.W. Blassingame (Baton Rouge: Louisiana State University Press, 1977):340.

Digital Collections

Cancer Alley Coalition accessed at https://www.enddeathalley.org.

Claire Carew website accessed at https://www.clairecarew.com/index.php?vw=1024&vh=1024&v0=&v1=128&v1b=0&v2=can&v3=0&v25=977.

Digital Public Library of America accessed at http://dp.la/primary-source-sets/exodusters-african-american-migration-to-the-great-plains.

Documenting the American South accessed at https://docsouth.unc.eduEcowatch accessed at https://www.ecowatch.com.

Fazendeville at WGNO accessed at https://wgno.com/news-with-a-twist/culture/the-story-of-fazendeville-an-african-american-town-in-the-middle-of-a-battlefield/

Goldman Prize Foundation accessed at https://www.goldmanprize.org.

Hurricane Digital Memory site accessed at http://hurricanearchive.org/.

Kansas Memory, Kansas Historical Society accessed at https://www.kshs.org/index.php?url=km/items/view/218451.

Library of Congress African American History Collections accessed at https://www.loc.gov/collections/?fa=subject:african+american+history.

Louisiana Bucket Brigade accessed at https://labucketbrigade.org/.

"Louisiana Slave Conspiracies" accessed at Https://lsc.berkeley.edu/1795.

Lower Ninth Ward Living Museum accessed at http://www.19livingmuseum.org/.

Mary Wheeler Collection, McCracken County Public Library, Digital Collections accessed at https://digitalcollections.mclib.net/luna/servlet/detail/McCracken~13~13~129~2241.

Mississippi Encyclopedia, Levee Camps accessed at http://mississippiencyclopedia.org/entries/levee-camps/.

MyNewOrleans.com accessed at https://www.myneworleans.com.

National Archives African American Collections accessed at https://www.archives.gov/research/african-americansNAACP Records, Manuscript Division, Library of Congress accessed at https://www.loc.gov/exhibits/naacp/the-great-depression.html.

New Orleans, Louisiana accessed at https://www.nola.com/.

Poor People's Campaign accessed at https://www.poorpeoplescampaign.org.

Resilience at https://www.resilience.org/.

RISE St. James accessed at https://earthjustice.org/features/cancer-alley-rises-up/.

Ted Ellis website at https://tedellisart.com/.

United Nations International Panel on Climate Change at https://www.ipcc.ch/reports/.

University Network for Human Rights accessed at https://www.humanrightsnetwork.org.

Yusef Komunyakaa website at https://www.poetryfoundation.org/poets/yusef-komunyakaa.

Discography

Barbecue Bob (1927) "Heavy Water Blues"

Bessie Smith (1927) "Backwater Blues"

Bessie Smith (1927) "Homeless Blues"

Big Bill Broonzy (1937) "Southern Flood Blues"

Charley Patton (1929) "High Water Everywhere"

Gene Campbell (1930) "Levee Camp Man Blues"

Lonnie Johnson (1928) "Broken Levee Blues"
Lucille Brogan (1927) "Levee Blues"
Memphis Joe and Kansas Minnie (1929) "When the Levee Breaks"
Son House (1941) "Levee Camp Moan"
Washboard Sam (1941) "Levee Camp Blues"

INDEX

Note: *Italic* page numbers refer to figures.

For Product Safety Concerns and Information please contact our EU representative GPSR@taylorandfrancis.com Taylor & Francis Verlag GmbH, Kaufingerstraße 24, 80331 München, Germany